a+u

《a+u》中文版 全新全线阵容搭建！

拳拳出发之际，我们搜寻逐梦的同行者、建筑的发声者！

招募职位

副主编：
有5年以上媒体主编或副主编担当经验；
熟悉平面媒体的管理模式，能够制定和把握刊物的整体方向；
熟悉建筑及相关行业领域的媒体定位和运作，有较丰富的行业及媒体资源者优先；
有良好的团队精神和组织沟通协调力、栏目策划力；
热爱媒体事业，具备良好的敬业精神和职业操守。

资深编辑：
有3年以上书刊工作经验，具备独立或者带领团队操作项目的经验者优先；
建筑、城市规划等相关专业本科或以上学历，拥有较高的编辑素养和敬业精神；
了解国家出版的相关法律法规，熟悉出版编辑规则及系列工作流程；
具有扎实的文字功底，较强的专题策划能力、决断力、判断力和沟通能力者优先；
对外联络和沟通能力强，做事细心周全，具有临场应变能力和较强的活动执行能力者优先；
热爱新媒体，具备良好的信息敏感度。

编辑：
大学及以上学历，中文、新闻、建筑等相关专业，有工作经验者优先；
思维活跃，有较强的文字编辑能力；拥有较高的专业素养和信息敏感度；
踏实勤奋，善于学习和接受新鲜事物，工作细致，责任心和沟通力强，具有团队合作精神；
可熟练使用英文或日文；可操作InDesign，Illustrator，Photoshop等软件者优先。

美术编辑：
1–2年工作经验；有较扎实的书籍装帧设计基础，熟悉图书出版工作流程者优先；
精通设计，灵活运用专业排版和设计软件：InDesign，Illustrator，Photoshop等；
熟悉视频制作或者剪辑等软件；审美创意一流，具有深厚的美术功底和创意构思能力者优先。

中英/中日翻译：
（兼职、长期招募）
熟悉建筑、城市规划类词汇；有扎实的中外文翻译功底（中文优秀者优先）；
建筑、城市规划等专业相关背景者优先；细心认真，有责任心，能在约定的时间内高质量完成工作。

欢迎有志之士加入，请将您的简历投至邮箱：hr@cagroup.cn

Architecture and Urbanism
Chinese Edition
2016:06 No.064

A+U 出版社

日文版
发行人：
吉田信之

客座编辑：
马卫东

编辑：
张替安佐子 郭莲明
池田绘里佳 服部真吏

设计顾问：
麦西莫·维格奈里

中文版
发行人／主编：
马卫东

编辑：
完颖 吴瑞香
王梦佳 巫盈盈

《建筑与都市》中文版编辑部
上海市大连路970号706室
(邮编：200092 电话：+86-21-33773001)

图书在版编目(CIP)数据

建筑与都市.百家争鸣：汉英对照/《建筑与都市》中文版编辑部编.--上海：同济大学出版社,2016.6
ISBN 978-7-5608-6361-0

Ⅰ.①建… Ⅱ.①建… Ⅲ.①城市建筑—建筑设计—作品集—世界—现代 Ⅳ.①TU984
中国版本图书馆CIP数据核字(2016)第121334号

建筑与都市 百家争鸣
《建筑与都市》中文版编辑部 编

出 品 人：华春荣
责任编辑：胡毅
助理编辑：李杰
责任校对：徐春莲

出版发行：同济大学出版社 www.tongjipress.com.cn
地 址：上海市四平路1239号（邮编：200092 电话：+86-21-65985622）
经 销：全国各地新华书店、建筑书店、网络书店
印 刷：上海雅昌艺术印刷有限公司
开 本：889mm × 1194 mm 1/16
印 张：10.5
字 数：336千字
版 次：2016年6月 第1版 2016年6月 第1次印刷
书 号：ISBN 978-7-5608-6361-0
定 价：98.00元

特辑：百家争鸣

Architecture and Urbanism
Chinese Edition
2016:06 No.064

A+U Publishing Co,. Ltd.

Japanese Edition
Publisher:
Nobuyuki Yoshida

Guest Editor:
Ma Weidong

Editorial Committee:
Asako Harikae Guo Lianming
Erika Ikeda Mari Hattori

Design Consultant:
Massimo Vignelli

Distributor:
Shinkenchiku-sha Co., Ltd.

Chinese Edition
Publisher / Chief Editor:
Ma Weidong

Editorial Committee:
Wan Ying Wu Ruixiang
Wang Mengjia Wu Yingying

Tongji University Press

Producer:
Hua Chunrong

Executive Editor:
Hu Yi
Assistant Editor:
Li Jie
Executive Proofreader:
Xu Chunlian

Cover: Image of architecture and cities. Image by a+u.

Editorial Department: CA-GROUP (Shanghai)
Address: Room 706, No. 970 Dalian Rd. Shanghai
(Hi-Shanghai Loft #9), 200092 China
Tel: +86-21-33773001 Fax: +86-21-33773336
http: //www.cagroup.cn E-mail: editor@cagroup.cn

Publisher: Tongji University Press
Address: 1239 Siping Rd. Shanghai, 200092 China
Tel: +86-21-65985622
http://www.tongjipress.com.cn

Printed and bound in Shanghai by Shanghai Artron Colour Printing Co., Ltd.

Feature: Architects in China

The years since 2003 saw continuous rapid economic development in China. With its GDP surpassing developed countries, China became the second largest economy after the United States. The population in its two major cities, Shanghai and Beijing, leapt respectively to 24.26 million and 21.52 million, making the list of global metropolises. In urban construction, the urbanization rate rose from 40.53% to 54.77%. Provinces and municipalities actively responded to the national policy of "One Belt, One Road", prudently put forward by the central government, gradually shifting away from the surge of mass urban development and toward ideas like urban regeneration and quality urban life. The completion of 2008 Beijing Olympics, 2010 Shanghai Expo and Shanghai Disney Resort in 2016 brought the progress of Chinese architecture, which has recorded unprecedented achievements in design concepts and technology, into the global spotlight.

Looking back to the evolution over the last decade, there are two events that are significant to the history of contemporary Chinese architecture.

The first event is Wang Shu winning the 2012 Pritzker Architecture Prize. With the introduction of western culture, contemporary Chinese architecture is influenced by western theories, following the views of western architectural criticism. The following facts might provide a glimpse of this situation: a number of major national projects were designed by foreign architects; likewise, most architects active in Chinese architecture right now have studied or worked abroad. Although there are different opinions about Wang Shu's laureate, the first official reception of a Pritzker Architecture Prize in China provides a great opportunity for Chinese architects to rethink, evolve and more importantly, return to the investigation of Chinese culture that was once left aside. Projects do not necessarily need to reference western architecture. They can be approved and respected by international architecture criticism when they are made with national qualities and inspired by regional culture as well as individual background.

The other event is Ma Yansong winning the international competition for the Chicago Lucas Museum of Narrative Art in 2014. In 2006, Ma Yansong became the first Chinese architect to win an open international competition, successfully bidding for the Absolute Towers in Canada. After eight years, winning the bid for the Lucas Museum of Narrative Art set a new monument for Chinese architects. Success in bidding for an international cultural landmark is unprecedented for Chinese architects, especially competing against a number of well-known international architects and architectural offices, including Zaha Hadid. Chinese architecture truly went global.

These various setbacks and breakthroughs, along with the up and downs of the times, tempered Chinese architects, boosting their confidence. Their current and future place became clear, and a new era of Chinese architecture is surely being unveiled.

These various setbacks and breakthroughs along with the up and downs of the times drilled Chinese architects, boosting their confidence. Their place now and to be in became clear and the era of Chinese architecture is surely being unveiled by Chinese architects.

The title of this issue in Chinese, "百家争鸣" (Contention of a Hundred Schools of Thought) originally refers to the thriving debates of philosophers and scholars during the Spring and Autumn and Warring States period. Here, it is quoted to describe an image of free ideological collision in contemporary Chinese architecture. While "百花齐放" (Hundred Blossoms) (*a+u* 03:12) reflected the phenomenon of a variety of architecture emerging in China around 2003, "百家争鸣" (Contention of a Hundred Schools of Thought) reflects more profound architectural propositions. This issue features over twenty projects of Chinese architects, ranging from newly built to various kinds of transformation projects. These projects, some of which are located in metropolises like Beijing, Shanghai and Guangzhou and others in the vast countryside, include a selection from large cultural institutions to mobile space covering only a couple of square meters.

Chief Editor, Ma Weidong

自2003年以来，中国经济持续快速增长，国民生产总值赶超发达国家，成为仅次于美国的世界第二大经济体。两大代表城市——上海和北京，人口规模分别膨胀至2,426万和2,152万人，一跃跻身“全球超级大城市”之列。在城市建设方面，全国城镇化率也从原来的40.53%增至54.77%。在中央政府提出“一带一路”的国家发展策略的同时，各地政府积极践行、审慎发展，从“城市大开发”的建设浪潮中激流勇退，逐步转向以“城市更新”、“提升城市生活品质”为理念的开发。从2008年北京奥运会、2010年上海世博会的相继举办，到2016年上海迪士尼的建成，中国建筑的发展令世界瞩目，无论是设计理念还是技术手段都达到了一个前所未有的高度。

纵观这十几年的变化发展，有两件事情对当代中国建筑史有深刻的意义。

一是2012年王澍获得普利兹克建筑奖。西学东渐，当代中国建筑受西方思潮影响，常以西方建筑为价值取向，比如之前中国的多个重大项目都由国外建筑师担纲，以及现在活跃在中国建筑界的建筑师大多拥有海外留学和工作的背景，我们从中可窥一斑。王澍获奖，虽然坊间对此说法各异，但普利兹克建筑奖首度落地中国本土，对中国建筑师来说，是思考更是发展的契机——重新审视被冷落的中华文化。将作品根植于地域文化底蕴，立足在自有背景，坚持民族品质，不必嫁接西方，同样会得到国际建筑界的尊重与认可。

二是2014年马岩松赢得芝加哥卢卡斯叙事博物馆的国际竞赛。2006年马岩松中标加拿大“梦露大厦”，这是中国建筑师首次赢得海外的公开国际竞赛。时隔8年，卢卡斯叙事博物馆的竞标成功，为中国建筑的发展立下新的里程碑——首次赢得了国际标志性文化建筑的设计权，尤其是在与包括扎哈·哈迪德等多位国际著名建筑师或事务所的角逐之后。这实现了中国建筑师真正意义上的“走出去”。

随着时代的起伏，大大小小的突破和挫折，磨练了中国建筑师，也极大地增强了他们的自信，使他们认清了自己所处和应处的国际坐标。相信中国建筑师正在为世界开启中国建筑的时代。

本辑的标题“百家争鸣”，原指春秋战国时期诸子百家各学派竞相争鸣的繁荣气象。这里大胆借取其意，表现当代中国建筑各种理念思想相互碰撞的自由景象。如果说之前的“百花齐放”(《a+u》03:12)映射的是2003年前后各类建筑开始在中国自由地呈现，那么“百家争鸣”反映的则是比“呈现”更具思想内涵的建筑主张。本辑选取了20余个中国建筑师的作品，它们有的位于大都市如北上广，有的在广袤的乡村；有新建的项目，也有改建增建的项目；有大型文化设施，也有几米见方的小型移动空间，这些作品相鸣争艳，形成中国建筑的新画面。从“百花齐放”到“百家争鸣”，中国建筑步履不歇，下一个10年，中国建筑又将呈现怎样的景象，不可想象，亦可想象。

主编　马卫东

Read the Architecture in China by Graphs

从图表看中国建筑

The last 10 years have seen tremendous economic expansion in China. By 2014 the gross domestic product had increased to almost four times over the Gross Domestic Product (GDP) of 2004. In 2004, China ranked seventh in GDP in the world, but by 2014 it had become the second largest economy, just after the U.S. The last decade has also been a period marked by mass urbanization as well as an explosion of real estate development and construction in China. Construction output made up 18% of the GDP in 2004. In 2014, it accounted for 27.7 % of the GDP. The growth in construction and real estate provided great opportunities for the progress of architecture. Hence, it is essential to closely examine the indicators in economics, construction and architecture to have a thorough understanding of the progress of architecture in China in the last decade. Graphs of key indicators are presented to reflect social, economic and architectural trends in China from 2004 to 2014. A comparison of major economic indicators of the two largest major cities, Beijing and Shanghai, is also provided as a data reference for the theme of this issue.

中国的经济在过去的10年间发展迅速，GDP在2004年到2014年间增长了将近4倍，从世界第7位一跃成为仅次于美国的世界第二大经济体。过去的10年也是中国城市化、房地产业及建筑业迅猛开拓的一段时期，建筑业在国内生产总值中的占比从18%上升到了27.7%，为中国建筑设计业的发展提供了良好的环境。因此，我们须从中国的经济、建筑业及建筑设计行业等方面的增长变化，来理解中国建筑在过去10年内的发展。本章所列的图表展示了2004年到2014年间中国社会经济、建筑业和建筑设计业的发展趋势。在此基础上，另选取了北京和上海这两座最具代表性的中国城市来进行比较，为本书的主题提供数据参考。

Notes

All graphs on pp. 6–7 are based on the data collected from National Bureau of Statistics of China, Beijing Municipal Bureau of Statistics, Shanghai Municipal Bureau of Statistics.

注：

第6–7页所列图表中的统计数据均来自中国国家统计局、北京市统计局和上海市统计局。

Major Social and Economic Indicators (2004–2014)

主要社会经济指标（2004–2014年）

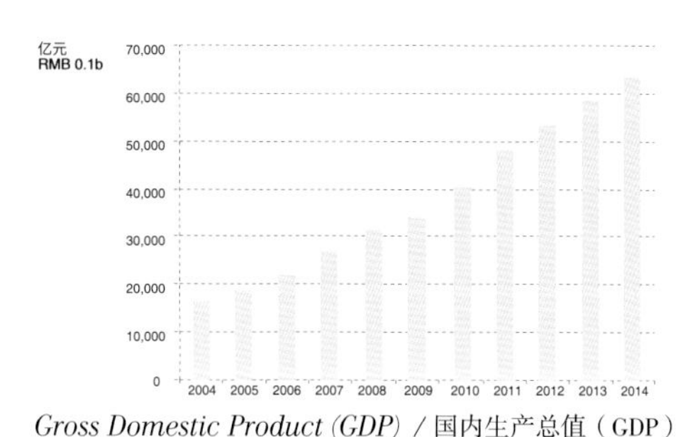

Gross Domestic Product (GDP) / 国内生产总值（GDP）

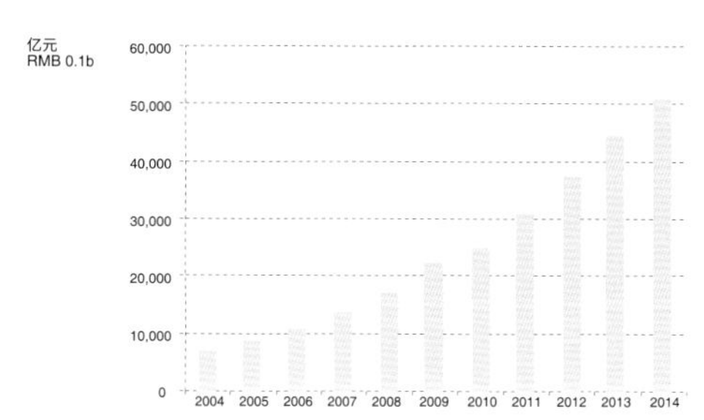

Fixed asset investment / 固定资产投资额

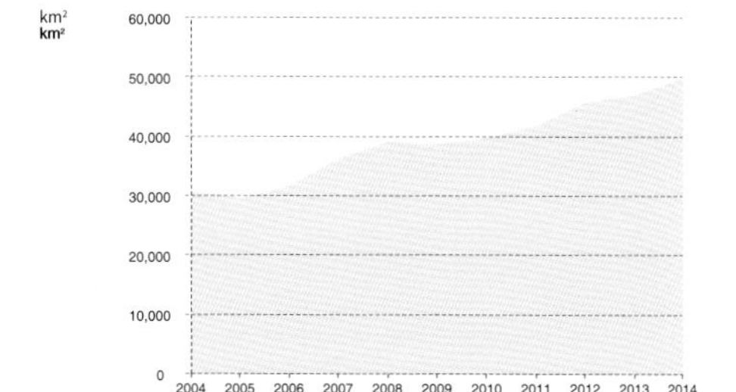

Urban construction area / 城市建设用地面积

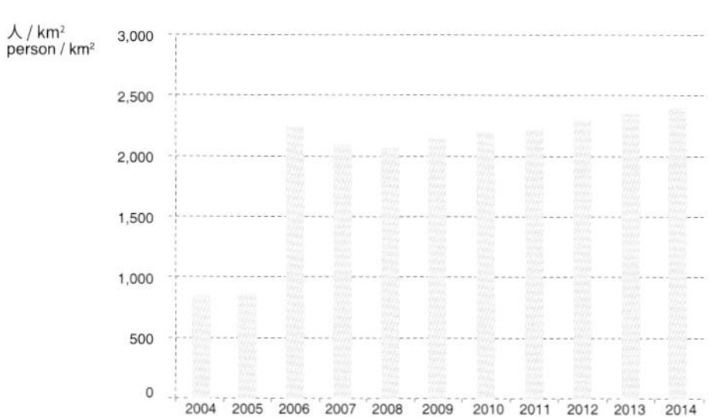

Urban population density / 城市人口密度

Construction Economic Indicators (2004–2014)
建筑业经济指标（2004–2014年）

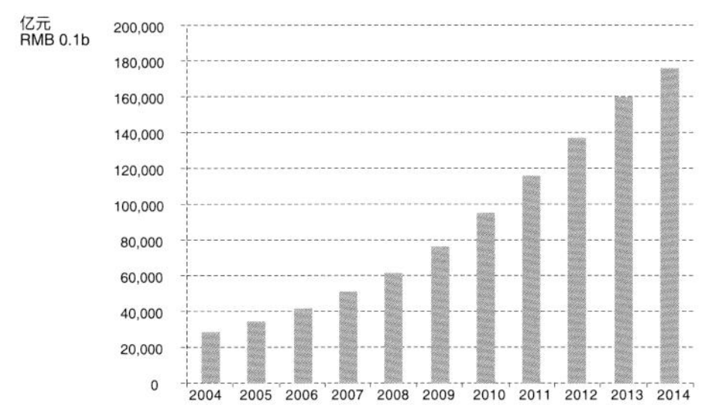

Total output value of construction / 建筑业总产值

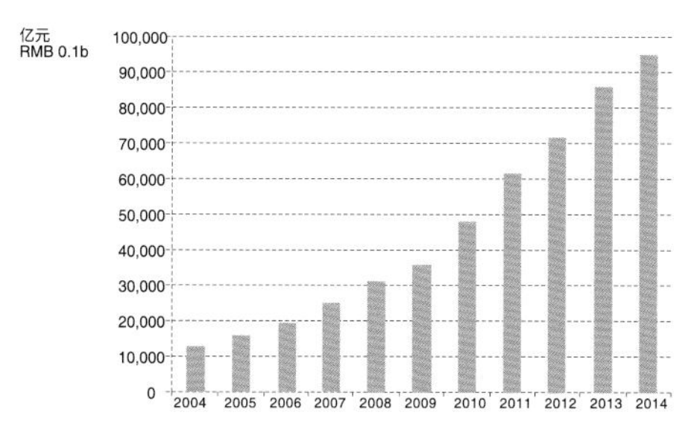

Investment in real estate development / 房地产开发投资额

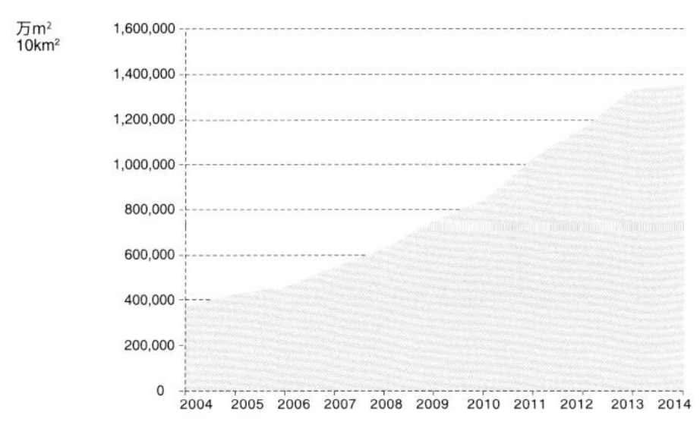

Construction floor area / 房屋施工面积

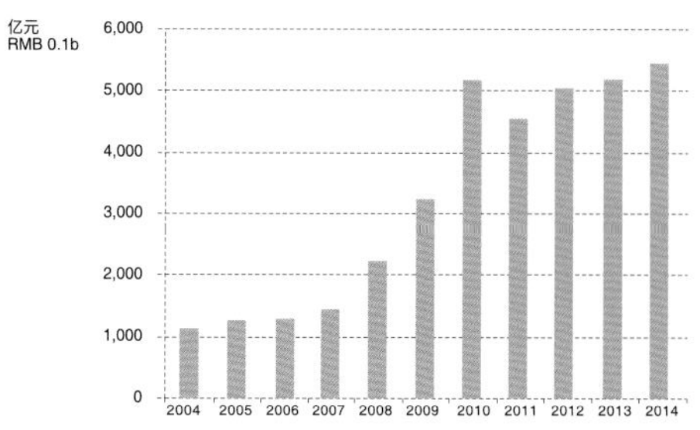

Investment in urban infrastructure / 城市环境基础设施建设投资额

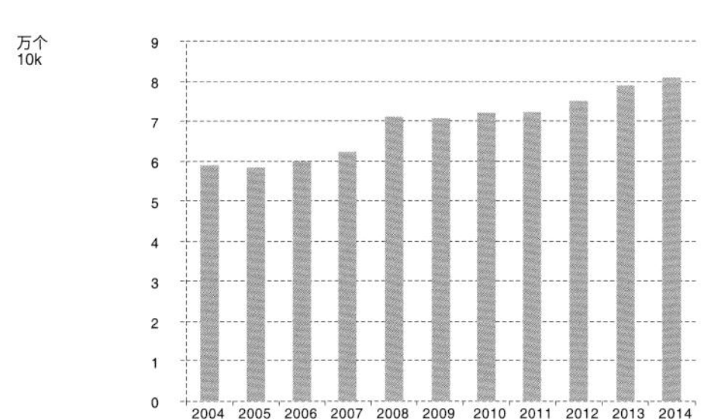

Number of construction enterprises / 建筑业企业单位数

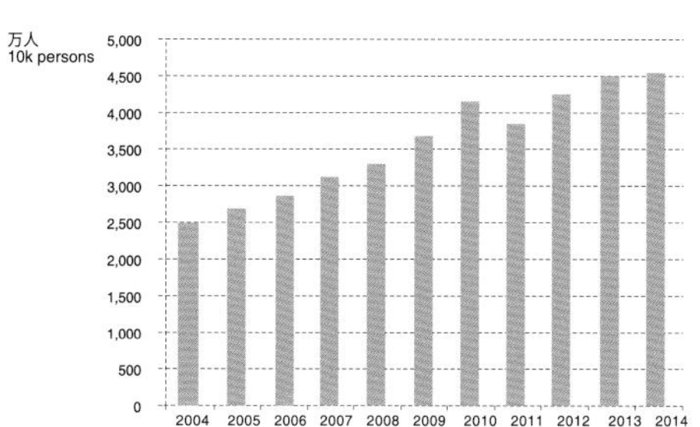

Persons employed in construction enterprises / 建筑业企业从业人数

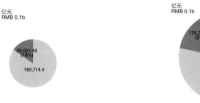

Share of construction output in GDP, 2004 / 2004年建筑业GDP份额

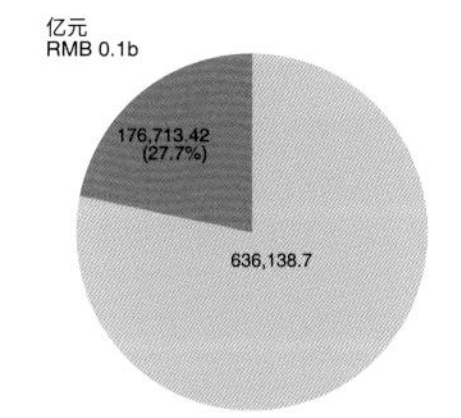

Share of construction output in GDP, 2014 / 2014年建筑业GDP份额

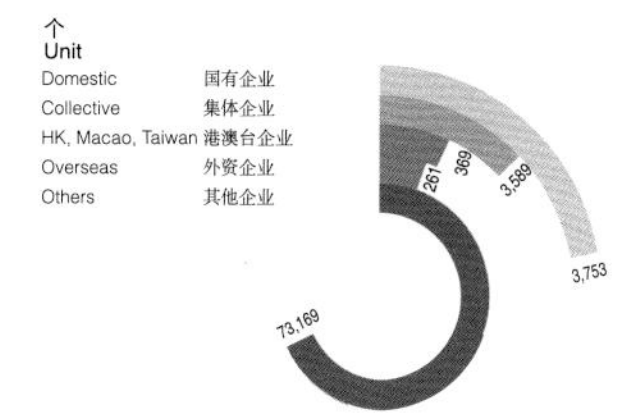

Number of construction enterprise by category, 2014 / 2014年建筑业企业分类

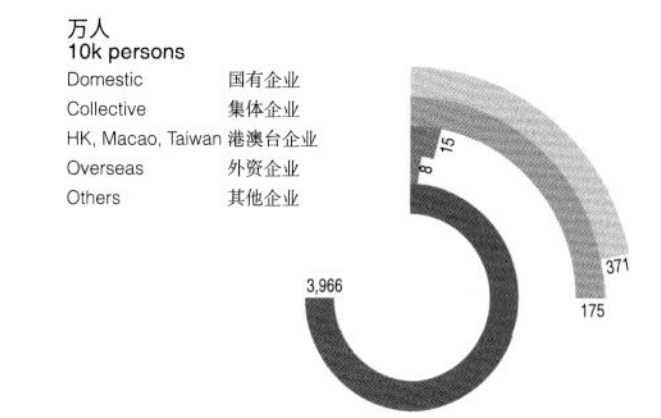

Employment distribution in construction enterprises, 2014
2014年建筑业企业从业人员分布

Comparison of Socioeconomic Indicators between Beijing and Shanghai (2014)
北京与上海的主要社会经济指标比较（2014年）

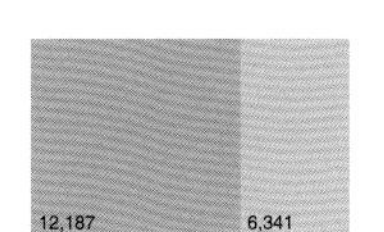

Urban area (km^2)
城区面积（km^2）

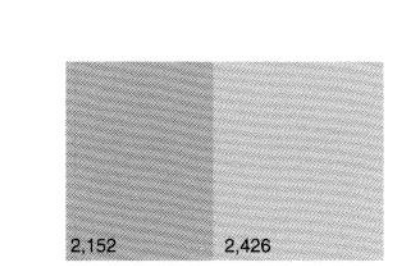

Population (10,000 persons) / 常住人口（万人）

Registered population (10,000 persons) / 户籍常住人口（万人）

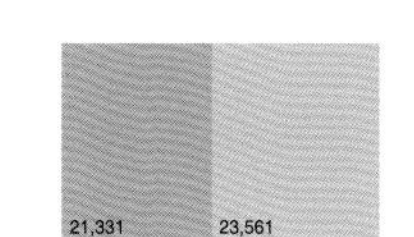

Total GDP (RMB 0.1b)
生产总值（亿元）

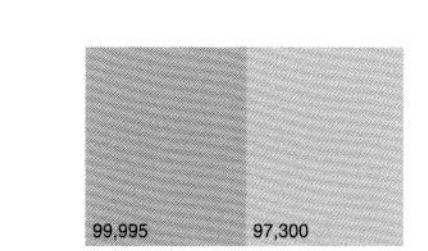

GDP per capita (RMB)
人均生产总值（元）

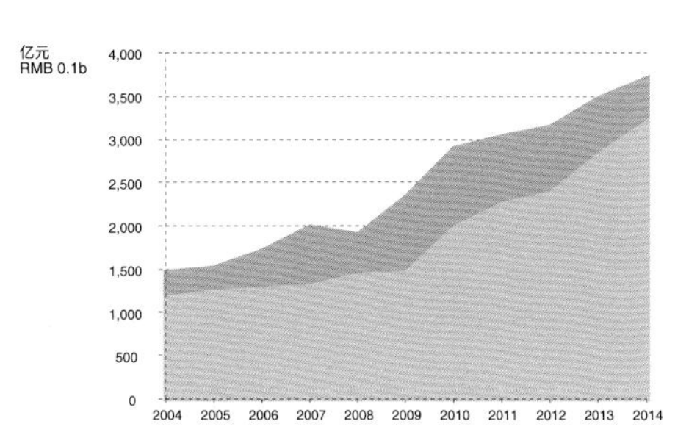

Investment in real estate development
房地产开发投资额

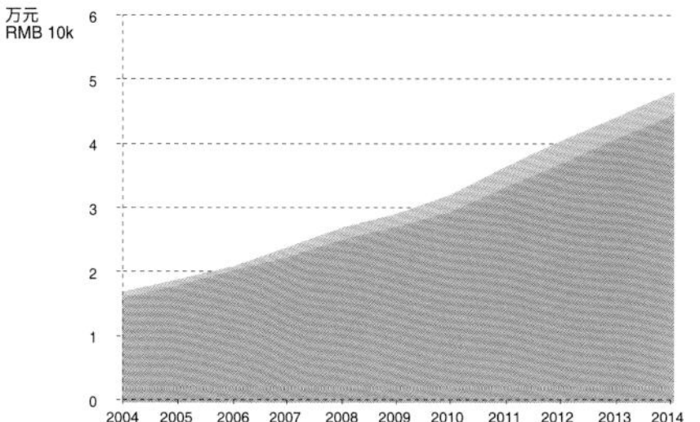

Disposable income of urban residents per capita
城镇居民人均可支配收入

Essay:

From "Experimental Architecture" to "Critical Pragmatism": Contemporary Architecture in China

Li Xiangning

论文

从"实验建筑"到"批判的实用主义"：论中国当代建筑

李翔宁

王梦佳 译

Since the establishment of the New China in 1949, the key players among Chinese architectural practices have always been the state-owned medium and large design institutes. Privately operated architectural firms began to emerge in the1980s and arguably, they started to draw public attention as practices independent of state ownership since Yung Ho Chang founded Atelier FCJZ in 1993. The discussion about experimental architecture among architectural journals is in effect attention paid to the practice of the then budding individual architects and an experimental paradigm, different from that of the mainstream state-owned architecture institutes, which emerged from the practices of these rising star architects.

Different from the experimental architectural practices in the 1990s, today there are a larger number of younger individual architects with more diversified patterns of practice. Without evident common guiding principles or revolutionary commitment, they work for the government, private developers, small private owners and various other sorts of clients, and more flexible and adaptive strategies have been evolved. "Critical pragmatism" may be an appropriate term to describe the collective practices of contemporary Chinese individual architects of the new millennium.

1. Experimental Architecture as Resistance

No matter whether the term "experimental architecture" is agreed upon or not, it is indisputable that in the 1990s a kind of "new" architectural practice "different" from the past practice appeared in the Chinese architecture. Such "new" and "different" aspects are not only reflected in the architecture image, but also in the more independent way of thinking and manner of working. Many scholars in China described this "new" architectural practice as "experimental architecture", and accordingly the designers, most of whom were emerging individual architects, were tagged as "experimental architects". Over ten years, these "experimental architects" have turned into a group of star architects with vigorous creativity and outstanding works executed now and then in contemporary China.

As a matter of fact, the so-called "experimental architecture" and "experimental architects" were closely bound up to the specific social, economic, and cultural backdrop of China in the late 1980s and early 1990s. At the time, a new trend was surging in both contemporary art and architecture, in the desire to break through the tradition and communicate with the world. The development of contemporary art and the introduction and translation of foreign architectural theories set the stage for Chinese "experimental architecture". Yung Ho Chang, Ma Qingyun, Wang Shu, Liu Jiakun, and Tang Hua were definitely the leading figures among the Chinese "experimental architects".

Apart from architects and their works, a series of architecture events (exhibitions, forums, etc.) evolved around Chinese "experimental architecture". Although it might be difficult to clearly define "experimental architecture" in China, a rough list of related key words will help understand this concept: young architects, individual practice, privately operated architectural firms, avant-garde, marginality, contemporariness, Chineseness, etc. As Rao Xiaojun pointed out in his article, "express a strong questioning attitude and challenge gesture towards orthodox or mainstream architecture trends and ideas from the very beginning". I myself would rather use the term "resistance" or "refusal", to describe the attitudes of the independent individual architects, to use architecture as vehicle to challenge the Western and Chinese mainstream architectural discourse and ideology, with or without consciousness in their pursuit of avant-gardism.

2. Diversified Architectural Practice and Critical Pragmatism

Today's China has witnessed individual architects establishing their own tribes and clusters in several important cities. These individual architects meet and communicate in various publications, exhibitions, collective design activities and seminars, revealing different subpopulation characteristics of different cities. In addition, they participate in the architectural education at well-known Chinese architectural schools in small groups. With a focus on architectural practice, these architects also develop their career in multiple dimensions including exhibitions, writing, teaching, academic research, and cultural communication. Such a state is quite different from that of Western architects and also their predecessor Chinese architects, and even those in the contemporary state-owned design institutes. Confronting the complex social and cultural environment and featuring different living conditions and practice strategies, they are seeking their personal positions in such an intricate relation network of different poles such as architectural autonomy vs. social reality, globalization vs. localization, politics vs. form.

Firstly, they are confronted with the relation between architectural autonomy and social reality. Architecture has always been struggling between its autonomy and reliance on capital and politics. The contemporary architecture discipline is unable to ignore the social economy and political culture to realize its pure autonomy. Therefore the majority of current

自1949年新中国成立以来，中国建筑界一直由国有大中型建筑设计研究院主导。民营建筑事务所的出现和发展在20世纪80年代之后，自张永和在1993年成立“非常建筑”（Atelier FCJZ）起，这一类独立于国有制之外的建筑单位就开始受到公众的关注。建筑期刊中对“实验建筑”的探讨，事实上是对新生代独立建筑师及其展现出的有别于主流国有建筑院的实验性范例的关注。

与90年代的实验建筑实践相比，今天很多年轻的独立建筑师有更多样的实践方式。他们没有明显普适的指导原则，也没有革命性的承诺，只是通过更加灵活和适应性更强的策略，来为政府、开发商和中小型企业等不同的业主主体服务。因此，如果要描述新世纪的当代中国独立建筑师的实践，使用“批判的实用主义”可能更为合适。

1. 抵抗的实验建筑

无论“实验建筑”这个术语是否被认可，无可非议的是90年代的“新”建筑实践与过去中国涌现出的建筑“不同”。这种“新”与“不同”不仅反映在建筑形象上，还表现为一种更为独立的思维与工作方式。中国的很多学者形容这种新型建筑实践为“实验建筑”，并相应地赋予这一类设计者们“实验建筑师”的称谓，而他们其中多数为独立建筑师。经过十多年的时间，这些“实验建筑师”成长为中国当代的明星建筑师，他们具有丰富的创造力，并且时常会设计出杰出的作品。

所谓 “实验建筑”和“实验建筑师”，都与中国20世纪80年代末到90年代初特定的社会、经济和文化背景密切相关。在这一时期，一种新的、想要打破传统与世界交流的思潮在当代艺术界和建筑界中蔓延开来。随着现代艺术的发展和国外建筑理论相关译著的传入，“实验建筑”应运而生，而张永和、马清运、王澍、刘家琨和汤桦正是这些“实验建筑师”中的领军人物。

除了建筑师和他们的作品，还有一系列的建筑活动（展览、论坛等）也与中国的“实验建筑”有关。尽管我们很难明确地定义中国的“实验建筑”，但以下这些关键词也许可以帮助我们更好地理解这个概念：青年建筑师、独立实践、个人建筑事务所、先锋、边缘性、当代、中国性……正如饶小军在他的文章中所指出的：“（独立建筑师）表达出一种强烈的质疑态度，并且从最开始就展现出挑战正统主流建筑趋势和想法的姿态。”我个人更偏向使用“抵抗”而非“拒绝”来形容他们的态度——一种在追求先锋主义的过程中有意无意地将建筑作为一个媒介，去挑战中西方主流建筑的话语与意识形态。

2. 多样化的建筑实践和批判的实用主义

现今的独立建筑师们在中国几个主要城市建立起了他们自己的聚集地，并通过各种出版、展览、合作设计活动以及研讨会等形式进行直接交流，这些都使处于不同城市的集体呈现出不同的集体性格。此外，他们也以小团队的形式参与中国著名建筑院校的建筑学教育工作；同时也在专注于建筑实践的基础上，广泛涉猎展览、写作、教学、学术研究和文化交流等领域，多元发展建筑师这一职业。这种职业状态与西方建筑师、早期的中国建筑师，甚至是同一时期在国有设计院工作的同行有着极大的区别。面对复杂的社会人文背景、不同的生活环境和实践策略，他们在一个极端且错综复杂的关系网中找寻自己的位置，这个关系网既形成于建筑的自主性与社会现实之间，也形成于全球化与本土化之间、政策与形式之间。

首先，他们面对的是建筑的自主性与社会现实之间的关系。建筑总是在其自主性和对资金、政策的依赖关系之间挣扎。当代建筑在实现其纯粹的自主性过程中必须兼顾社会经济与政治文化，因此，当今大部分的独立建筑师都直面现实，并试图在与现实协同的过程中找到一个可以最大化利用现实的平衡点，以此为自己争取一个批判性的立场。政府、开发商、文化学会、私人业主等业主类型的多样化，以及项目性质、尺度的持续变化等因素也许解释了这样一个问题——为什么建筑师总是在不断地改变自己的设计策略，而鲜有建筑师可以维持一种一贯相承的建筑语言。这种生存技能多多少少是一种被动的选择，也折射出了当代独立建筑师必须面对的现实。

其次，他们须面对全球化和本土化之间的关系。海外的教育经历给很多中国当代青年建筑师们带来了国际化的经验和视野，使他们敏锐地觉察并审慎考虑全球化和本土化之间相互对立的关系，以及如何以当代的方式解读“中国性”的问题。他们用一种国际化、现代化的形式去回应本土的环境，这也许是一种协商和权宜的结果，却很好地传达了建筑师的设计理念与复杂的社会现实之间的平衡需求。在中国独立建筑师中，有张永和、马清运等拥有海外教育背景的建筑师，在国内进行着建筑实践，同时担任建筑学院国际建筑教育的院长或领导

individual architects now face up to the reality and try to achieve a critical position through cooperation with reality and to make the most of it. Diversity in client types, such as the government, real estate developers, cultural institutes, and individual proprietors, and constant variability in project nature and scale may explain why few architects maintain a complete and persistent architecture language, but constantly change strategies. Such survival guidance more or less coming out of passive choice, might have promulgated the reality that many contemporary individual architects have to be confronted with.

Secondly, they are confronted with the relation between globalization and localization. Thanks to the international experience and vision brought by their overseas education, many young contemporary Chinese architects are sensitive enough to take the initiative to think over the position of architectural practice in the binary opposition of globalization and localization, as well as how to interpret issues of "Chineseness" in a contemporary manner. To respond to local conditions with a global modern form may be a result of negotiation and makeshift, but it has commendably presented the balance between the architects' design ideal and the complicated social reality. Among Chinese individual architects, we can find those who preside over international architectural education as deans and heads of architecture schools with their oversea education background and architectural practice at home, such as Yung Ho Chang and Ma Qingyun, those who win international building projects by constant participation in international competitions like Ma Yansong, and those who are educated in China but internationally accredited through persistent exploration of "Chinese contemporary" architectural identity like Wang Shu.

Finally, they have to achieve a balance between politics and form. Form, as a matter of fact, is an important part of the core of the architectural discipline. No matter how much an architect claims his works are divorced from formalism, form has taken root in his architecture. The larger system of contemporary Chinese architecture has developed competition evaluation criteria based on judgment of form in most competitions and tendering. Whatever approaches the evaluation takes, what is to be resolved in the end is a proposition of form. The trajectory of contemporary Chinese politics has inevitable influence on architectural form. City policy makers with obscure political desires and aesthetic tastes may eventually decide the formal criteria of architecture and urban design. However, politics can also have positive influence on architecture and urban design given proper guidance.

3. Current Opportunities and Challenges

Interaction between the traditional core of the architectural discipline and new technological innovation has provided new chances to review the practice of contemporary Chinese individual architects and therefore look for the way for future development and the possibility of making a special contribution to architecture. What we now observe from contemporary Chinese architecture is mostly variations of typical Western modern architecture, regarding space, dimension, structure, material, function, or if going a little further, light, poetics, and tectonics. From the practice of these individual architects, it is hard to find an independent and insistent architectural language (which Wang Shu's works may have), or mature strategies that are responsive to the rapid transformation taking place in contemporary Chinese cities, as well as continued attempts in architecture innovation and industrialization. But still some trends could be noticed in the practice of individual architects and these may become driving factors for the future Chinese architectural evolution in China.

Digital building technologies (including 3D printing, kinetic architecture, and BIM system) and international waves of parametric architecture brought by today's digital innovation have greatly influenced the practice of Chinese individual architects. Xu Weiguo, Philip F. Yuan, Song Gang and Wang Zhenfei are China's representational pioneers in this direction. The completion of some parametric buildings by those architects in recent years, such as the Silk Wall and Tea House in J-Office project built by Philip F. Yuan (Archi-Union Architects) in Shanghai for his workshop, and Yujiabao Engineering Control Centre by Wang Zhenfei, has demonstrated the possibility that digital building technology will take root and develop in China. Such an innovation does not only mean creative changes in fancy forms, but also changes in the architectural culture and way of space production. A recent example is the Phoenix International Media Centre designed by Shao Weiping, chief architect of the biggest state-owned design institute Beijing Institute of Architectural Design (BIAD).

Another trend developing simultaneously with digital architecture is the increasing concern about social justice and building ethics in the practice of individual architects. One symbolic event is the appraisal of the China Architecture Media Awards which since 2006 exert an influence on the public and the professional field. Advocating "civil architecture", this award places architectures' common and social character and contribution to society at a more important position than architecture aesthetics and the design itself. Award-winning

的建筑师；有马岩松这样通过不断参加国际竞赛而赢得国际建筑项目的建筑师；同时也有王澍这样在国内接受教育，但通过持续不断地探索“中国当代”的建筑本质而获得国际声誉的建筑师。

最后，他们还须维持政治与形式之间的平衡。形式，毋庸置疑是建筑学中很核心的一部分。无论建筑师如何声明自己的作品脱离了形式主义，形式本身仍根植于他的建筑中。中国当代的建筑体系发展出了一套基于形式、适用于大部分竞赛和竞标的评判标准。无论评判的方法是什么，最后都会归结到形式的命题上来。中国当代的政治轨迹不可避免地影响了建筑形式。城市决策者们含蓄的政治愿望与审美趣味会最终决定建筑与城市设计的正式标准，而如果政治方面的引导得当，将会对建筑和城市设计产生积极的影响。

3.当前的机遇与挑战

建筑的传统核心与新技术革新之间的互动，为我们重新审视当代中国独立建筑师们的实践提供了机会，从而探寻到未来建筑的发展之道以及在建筑领域作出特殊贡献的可能。我们现在观察到的中国当代建筑，无论从空间、尺度、结构、材料、功能上，或者更深远一点地从光、诗意、地质构造上看，更多的是典型西方现代主义建筑的变体。在这些独立建筑师的实践中，我们很难找到一种自成一格、历日弥久的建筑语言（也许在王澍的作品中出现了），一种可以回应当代中国城市快速变化的成熟策略，以及一种对建筑创新和工业化的持续尝试，但这些独立建筑师们的实践中仍然展现出一些值得注意的趋势，而且它们将可能成为未来中国建筑进程的推动力。

数字建筑技术（包括3D打印、动态建筑和BIM系统），以及由现今的数码创新带来的国际参数化建筑热潮，都对中国独立建筑师的实践有着深远的影响。徐卫国、袁烽、宋刚和王振飞是中国这方面代表性的开拓者。这些建筑师们近几年完成的一些参数化建筑，如袁烽为自己的工作室建造的“绸墙”和“五维茶室”，王振飞设计的“于家堡工程指挥中心”，都证明了数字建筑技术在中国扎根和发展的可能性。这不仅意味着各种外形上的创新型改变，还代表着建筑文化和空间创造方式的改变。最近关于这方面的一个项目是由邵卫平（中国最大的国有设计院——北京市建筑设计研究院有限公司的执行总建筑师）主持设计的“凤凰出版传媒总部”。

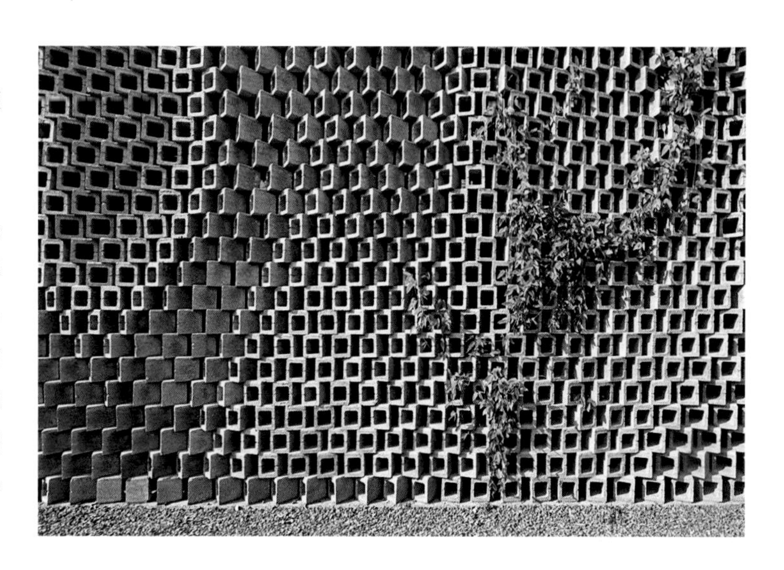

This page, above: Silk Wall by Philip F. Yuan. This project is conversion of a dilapidated warehouse into an architectural design studio. The concept of the Silk Wall (the external wall of the warehouse) was based on manipulating simple materials using up-to-date fabrication processes. This page, below: Bridge School by Li Xiaodong. The site is located in a traditional Hakka village. Since there are two Tulous separated by the creek at the centre of the village, the architect designed a primary school connecting them like a bridge over a creek. All photos on this page courtesy of the architects.

本页，上：袁烽的作品“绸墙”。该项目是将一栋废弃的工业老厂房改造成建筑设计工作室。老厂房的外墙“绸墙”设计基于平实的材料和最新的建造表现。本页，下：李晓东的作品“桥上书屋”。该项目位于一个传统的客家村落，村子被一条小河隔断，河两岸分别有两座土楼。这所小学被设计为一座横跨小河的桥廊建筑。

this award places architectures' common and social character and contribution to society at a more important position than architecture aesthetics and the design itself. Award-winning projects such as the Gehua Youth and Cultural Centre by Li Hu's OPEN Architecture, Maosi Ecological Experimental Primary School by Wu Enrong, Changxing TV Station by Fu Xiao, and the Xinya Primary Schools by Zhu Jingxiang have all pushed forward a kind of public culture value. More projects appear with higher sense of social responsibility like the Urban Tulou (See pp. 120–123) – affordable apartment for low-income group in cities designed by URBANUS adopts the form of traditional southern Hakka dwellings – ring-shaped earthen buildings. Inspiration from regional architectural style and user-friendly design approach are both mobilized to create nice form.

A new trend correlative to the increasing social responsibility and equity in architecture is that more individual architects turn their eyes to the countryside, design and construct country buildings and even take part in its social reconstruction. Until nearly ten years ago, architects have concentrated on the architecture design and construction in cities, ignoring the countryside because of its unfavorable economy and culture condition. While in recent years, the crowded streets and industrial pollution in cities have lost attraction to architects, who then turn to the countryside where the idyllic pastoral dreams still can be realized. There are projects like Bridge School in Fujian province designed by Li Xiaodong and serial rural construction including House for All Seasons (*a+u* 13:08) designed by the Hong Kong architect John Lin. As some artists, designers and social workers are stationed in the countryside and participate in community reconstruction there, some individual architects also persevere in building the countryside and even move out from their cities to become members of the local community, such as Wang Hao, Zhao Yang, Huang Yingwu, Chen Haoru, who have attracted a lot attention for their architecture activities in mainland China's countryside.

4. Conclusion

Today, the external environment for Chinese architects has been greatly improved over ten years ago. Even young architects are able to complete large-scale buildings with high construction quality. At the same time, the main challenge confronting them is how to choose among the ocean of styles, to hold their own positions and, and to keep the continuity of their formal identity and design strategy.

Although the individual architects have contributed a large number of good-quality works over decades, they have not responded forcefully to the distinctive characteristics of the contemporary Chinese urbanism such as bigness, swiftness, cheapness, uncertainty, and so on. Therefore, as we rejoice for the good buildings springing up through Chinese individual architects' efforts, we have to be aware that the improvement of design and construction quality also relies on the independent critical consciousness. Reflecting the ethos of our, paying more attention to social connotation of architecture, and exploring the latent influence of new technological innovation might be the impetus for the Chinese individual architects to step further with greater expectations.

Prof. Li Xiangning is deputy dean of history, theory and criticism at Tongji University College of Architecture and Urban Planning. He has published widely on contemporary architecture and urbanism in China and he is teaching a course on the same topic at the Harvard Graduate School of Design (GSD) in the year of 2016. His recent books include *The Real and the Imagined: Study of Value System in Contemporary Urban Theory* (2009) and *Updating China: Projects for a Sustainable Future* (2010).

与“数字建筑”这种趋势并势前行的是独立建筑师们对社会公正与建筑伦理愈发浓厚的关心。其中具有象征意义的事件是自2006年开始举办的“中国建筑传媒奖”，它对建筑业界内外都产生了影响。通过宣扬民用类建筑，该奖项将建筑的公众性和社会性，以及它们对社会的贡献，都放在了比建筑美学和设计本身更重要的位置上。其中获奖的李虎的OPEN建筑事务所设计的“歌华营地体验中心”、吴恩融的“毛寺生态实验小学”、傅筱的“长兴传媒中心”和朱竞翔的“新芽小学”，都推动了一种公众文化价值。与此同时，更多项目展现出了更高的社会责任感，如“土楼公舍”（详见第120–123页）——URBANUS都市实践参考南方传统的客家土制民居，为城市低收入群体设计的经济型公寓住房。地方建筑形式的启发、人性化的设计方法都有助于创造出优秀的建筑形式。

另一个与建筑界逐渐增强的社会责任感和公正感相关联的新趋势是更多的独立建筑师放眼乡村，设计并建造乡村建筑，甚至参与乡村的建设活动中。10年前，建筑师们还是将建筑设计和建造聚焦在城市，乡村因其不尽如人意的经济和文化现状而备受忽略。直到近些年，城市中拥挤的街道和工业污染使建筑师失去了对城市原有的热情，而转向乡村——在那里牧歌田野般的梦想仍有实现的可能。这其中有李晓东设计的福建省下石村的“桥上书屋”，以及香港建筑师林君翰设计的一系列类似“四季：一所房子”（《a+u》13:08）的项目。与一些驻守在乡村、参与当地社区建设的艺术家、设计师和社会工作者一样，一些独立建筑师坚持着乡村建设，甚至搬离城市，成为当地农村居民的一员，如王灏、赵扬、黄英武、陈浩如都因为他们在中国大陆农村的建筑活动而备受关注。

4.结论

在今天，中国建筑师们面临的外部环境比10年前进步了很多，甚至有些年轻的建筑师也有能力以很高的工程质量完成大型建筑。同时，他们面临着如何从浩瀚的建筑形式中，找定自己的位置并保持设计形式的统一性和设计策略的延续性等挑战。

尽管独立建筑师们几十年来完成了大量的高质量作品，但他们仍未对当代中国都市主义的诸如宏大、迅速、廉价、不确定性等显著特点作出有力的回应。因此，当我们为在中国独立建筑师们的努力之下所涌现出的优秀建筑作品而欣喜的同时，我们必须意识到设计和工程质量的进步仍须依赖独立的批判意识。我们民族精神的反映，对建筑的社会内涵的关注，对新的技术革新带来的潜在影响的探索，都将是中国独立建筑师们坚于道、行向前的动力。

李翔宁，同济大学建筑与城市规划学院副院长，历史、理论与评论的教授，发表了大量关于当代中国建筑与城市化理论和评论的文章，并于2016年在哈佛设计研究生院（GSD）讲授与此相关的课程。近期著作有《想象与真实——当代城市研究中价值视角分析》（2009）和《更新中国：为了一个可持续的未来》（2010）。

Atelier Deshaus was founded in Shanghai in 2001. Principal, Liu Yichun (left) was born in 1969, obtained Masters Degree from Tongji University, Department of Architecture in 1997. Principal, Chen Yifeng was born in 1972, obtained Master's Degree from Tongji University, Department of Architecture in 1998.
Deshaus attended major international exhibitions on contemporary Chinese Architecture in Shanghai, Beijing, Hong Kong, London, Paris, Tokyo, Vienna, Barcelona, Brussels, Prague, Venice, Milan, Rotterdam, Bordeaux, Cincinnati and Dusseldorf etc. In 2011, Atelier Deshaus was selected by the *Architectural Record* to be one of the ten firms in year's "Design Vanguard". In 2014, Atelier Deshaus awarded "the Architectural Review Emerging Architecture Awards" by the *Architectural Review*.

大舍建筑设计事务所于2001年在上海成立，主持建筑师柳亦春（左），生于1969年，于1997年毕业于同济大学建筑系，获建筑学硕士学位。主持建筑师陈屹峰，生于1972年，于1998年毕业于同济大学建筑系，获建筑学硕士学位。大舍参与了诸多关于中国当代建筑的重要国际展览，其作品在上海、北京、香港、伦敦、巴黎、东京、维也纳、巴塞罗那、布鲁塞尔、布拉格、威尼斯、米兰、鹿特丹、波尔多、辛辛那提和杜赛尔多夫等城市均有亮相。2011年，大舍入选美国《建筑实录》评出的年度全球十佳先锋设计事务所（Design Vanguard, 2011）。2014年，大舍获得由英国《建筑评论》颁发的全球新锐建筑奖（the Architectural Review Emerging Architecture Awards, 2014）。

Hua Li received his B. Arch. and M. Arch. from Tsinghua University, and then M. Arch from Yale University. Having practiced in New York, he founded **Trace Architecture Office** (**TAO**) in 2009 in Beijing. TAO explores the essence of place to make architecture deeply rooted in its cultural and environmental context with respect to local condition. Hua Li and TAO have won several important awards including the ARCASIA Award (2013), the *Architectural Record* "Good Design is Good Business Awards" (2012), and the WA Award (2012). In addition to his practice, Hua Li lectures extensively in both Asia and Europe, and teaches as a visiting professor at Tsinghua University. He has also been a guest critic for studio reviews at the Swiss Federal Institute of Technology (ETH) Zurich, The University of Hong Kong , The Central Academy of Fine Arts and the Berlin University of the Arts (UdK).

华黎，本科毕业于清华大学，并先后在清华大学和美国耶鲁大学建筑学院获得建筑学硕士学位，之后曾在纽约从事建筑实践。2009年，华黎在北京创立**迹・建筑事务所**（TAO）。TAO的实践通过深入挖掘场所意义和合理运用此时此地的条件，营造根植于当地文化与自然环境的建筑和景观。华黎以及TAO曾获得包括2013年阿卡汗国际建筑奖提名、美国《建筑实录》评选的2012年全球设计先锋以及最佳公共建筑奖、2012年 WA建筑奖在内的多个奖项。华黎曾受邀在亚洲和欧洲范围内的多项建筑活动中发表演讲，同时他作为清华大学的客座教授，教授设计课程，并曾受邀担任瑞士苏黎世联邦理工学院、香港大学、中央美术学院以及柏林艺术大学学生课程设计的客座评论员。

Zhang Lei was born in 1964, graduated from Southeast University in Nanjing and finished his postgraduate study in the Swiss Federal Institute of Technology (ETH) Zurich. He started his architecture practice AZL architects in 2001 in Nanjing. He is now teaching as professor in School of Architecture and Urban Planning in Nanjing University.
AZL was selected as one of the ten firms as "Design Vanguard" by *Architectural Record* in 2008. Concrete Slit House was entered for an honorable mention project by "the Architectural Review Emerging Architecture Awards" by *Architectural Review* in 2008. Zhang Lei's projects have been published extensively and involved in major international exhibitions on contemporary Chinese Architecture in the Netherlands, France, Germany, Italy and the US.

张雷，1964年出生，东南大学建筑系硕士，瑞士苏黎世联邦理工学院研究生。2001年创始张雷联合建筑事务所兼任主持建筑师，现任南京大学建筑与城市规划学院教授和可持续乡土建筑研究中心主任。
张雷联合建筑事务所在2008年入选美国《建筑实录》评出的年度全球十佳先锋设计事务所（Design Vanguard）。作品"混凝土缝之宅"荣获英国《建筑评论》2008年ar+d国际青年建筑师奖荣誉提名。张雷近年来的作品被国际高水平建筑杂志、刊物广泛发表收录，并参与了在荷兰、法国、德国、意大利、日本和美国等国家举办的诸多关于中国当代建筑的国际展览。

Dong Gong received Bachelor's and Master's of Architecture degrees from Tsinghua University and Master of Architecture degree from University of Illinois. He worked for Solomon Cordwell Buenz (SCB) in Chicago, Richard Meier & Partners Architects, and Steven Holl Architects in New York.
In 2008, Dong Gong founded **Vector Architects** and has become one of the most active young architects in China. Dong Gong has been invited as guest speaker by Tsinghua University, TianJin University and Southeast University, etc.. He is also teaching architecture studios at Tsinghua University. His firm and projects have won awards including China Architecture Media awards, WA Chinese Architecture Awards, Chinese Excellent Exploration and Design Industry Awards, Architecture and Design Trophy Awards, Blueprint Award, and *Architectural Record* "Design Vanguard".

董功，清华大学建筑学学士和硕士，美国伊利诺大学建筑学硕士，留美期间曾先后工作于美国芝加哥Solomon Cordwell Buenz & Associates, 纽约 Richard Meier & Partners和Steven Holl Architects。
2008年，董功创立**直向建筑事务所**，成为目前中国设计领域活跃的青年建筑师之一，曾多次受邀参与清华大学、天津大学、东南大学等国内外学府的演讲活动，并于2014年受聘担任清华大学建筑学院设计导师。直向建筑事务所及其作品曾多次获得国内外奖项，包括中国建筑传媒奖、WA第6届中国建筑奖佳作奖、全国优秀工程勘察设计行业奖一等奖、美国《建筑实录》杂志评选的2014年全球十佳先锋设计事务所（Design Vanguard）、2015 Blueprint Award最佳公共建筑类别特别推荐奖、2015 A&D Trophy Award机构/公共类别最佳建筑奖等。

Atelier Z+, established in 2002, is a Shanghai based interdisciplinary design team which covers urban design, architecture, interior and landscape. Atelier Z+ is pursuing an open and introspective working method instead of any specific form and style while carrying out tremendously distinguished projects. In the face of rapid and abrupt change of social value, material environment and life style in China, Atelier Z+ is persistent in treating the specific problems of the different projects as the starting point of work, aiming at integrating various resources into a highly consistent entirety. By revelation and presentation of the potentiality hidden in respective projects, we shall have an insight into various kinds of contradictions of the existing world: present and the history, local and global, environment and development, reality and ideality.

致正建筑工作室成立于2002年，是一个立足于上海的跨领域的设计实践团队，其工作涵盖城市、建筑、室内和景观设计，并在尺度差异巨大的不同项目中探讨一种不以特定形式风格为目标的、开放的、内省的工作方式。面对当下中国急速剧变的社会价值、物质环境和生活方式，致正建筑工作室始终以具体项目本身所面对的特殊问题为工作的起点，以将现实存在的、看似彼此孤立甚或冲突的各种资源整合成为一个内在高度一致的整体为目标，通过对于不同项目内在隐匿潜力的揭示和呈现，来洞察我们生活的世界的种种矛盾：当下与历史、本土与全球、环境与发展、现实与理想等。

Philip F. Yuan is a Professor in Architecture Department at Tongji University in Shanghai, and the director of the Digital Design Research Centre (DDRC) at the College of Architecture and Urban Planning (CAUP), Tongji University. He is also the founding director of **Archi-Union Architects**. As one of the founders of the Digital Architectural Design Association (DADA) of the Architectural Society of China (ASC), his research and practice focuses on digital design and fabrication methodology with the combination of Chinese traditional material and craftsmanship. His research publications include *A Tectonic Reality* (China Architecture & Building Press, 2011), as well as *Theatre Design* (2012), *Fabricating the Future* (2012), *Scripting the Future* (2012), *Digital Workshop in China* (2013) *Robotic Futures* (2015) and *From Diagrammatic Thinking to Digital Fabrication* (2016), all published by Tongji University Press.

袁烽，上海同济大学建筑与城规学院教授，同济大学建筑与城规学院数字设计研究中心的主管，他同时也是上海**创盟国际**建筑设计有限公司的创始人。作为中国建筑学会数字专业委员会（DADA）的联合发起人，他的研究和实践专注于数字化设计和建造方法与中国传统材料和工艺的结合。他的学术出版物包括《现实建构》（2011，中国建筑工业出版社），以及由同济大学出版社出版的《观演建筑设计》（2012），《建筑数字化建造》（2012），《建筑数字化编程》（2012），《探访中国数字建筑设计工作营》（2013），《建筑机器人建造》（2015）和《从图解思维到数字建造》（2016）。

MAD Architects, founded by Ma Yansong in 2004, is a global architecture firm committed to developing futuristic, organic, technologically advanced designs that embody a contemporary interpretation of the Eastern affinity for nature. With its core design philosophy of Shanshui City – a vision for the city of the future based in the spiritual and emotional needs of residents – MAD endeavors to create a balance between humanity, the city, and the environment. Founding principal Ma Yansong is a central figure in the worldwide dialogue on the future of architecture, and has been named one of the "10 Most Creative People in Architecture" by *Fast Company* in 2009, and selected as a "Young Global Leader (YGL)" by World Economic Forum (Davos Forum) in 2014. MAD is expanding its global presence with projects across the globe including Chicago's Lucas Museum of Narrative Art.

MAD建筑事务所由中国建筑师马岩松于2004年建立，是一所以东方自然体验为基础和出发点进行设计，致力于创造可持续并具未来感、有机的高科技建筑的国际建筑事务所。近年来围绕"山水城市"这一核心设计哲学，MAD期望通过创新建筑来建立社会、城市、环境和人之间的平衡。马岩松作为MAD的创始人兼合伙人，是未来建筑的世界性对话的重要代表，也是首位在海外赢得重要标志性建筑竞标的中国建筑师。2009年被*Fast Company*杂志评选为"全球建筑界最具创造力10人"之一，2014年被世界经济论坛评选为"世界青年领袖"。MAD通过系列作品如梦露大厦、哈尔滨大剧院、芝加哥卢卡斯叙事艺术博物馆等跨越全球的建筑项目在世界范围内实践着未来人居理想的宣言。

Zhu Pei received his Master's degree in Architecture both from Tsinghua University and the University of California at Berkeley. He founded **Studio Pei-Zhu** in Beijing in 2005. Zhu Pei was named one of "the 5 greatest architects under 50" by the *Huffington Post* in 2011, won the Courvoisier Design Award by *Wallpaper* 2009, "Design Vanguard" by *Architectural Record* in 2007, "China Award" from *Architectural Record* in 2005 and Design for Asia Awards for Grand and Culture Award, Hong Kong, in 2008, Special Merit Award by International Union of Architects (UIA) and UNESCO in 1989. His works have been exhibited at world important museums such as Venice Bienniale, Sao Paulo Biennial, Centre Pompidou, Victoria and Albert Museum.

朱锫，清华大学建筑学硕士，美国加州伯克利大学建筑与城市设计硕士。2005年创建**朱锫建筑设计事务所**。朱锫被美国赫芬顿邮报选为当今世界最重要的5位（50岁以下）建筑师之一（2011），被英国*Wallpaper*杂志授予库瓦西耶设计奖（2009），被美国《建筑实录》杂志评为全球设计先锋（2007），获中国建筑奖（2005），获香港DFA评选的亚洲最高荣誉设计大奖、亚洲文化优异设计大奖（2008），获国际建筑协会、联合国教科文组织评选的设计特别奖（1989），其作品先后在意大利威尼斯双年展、巴西圣保罗双年展、法国蓬皮杜艺术中心以及英国维多利亚博物馆等世界知名展览和美术馆中展出。

Wang Shuo, founding principal of **META-Project**, is an architect, researcher and curator based in Beijing. He received his B. Arch. from Tsinghua University in Beijing and M. Arch. from Rice University. He had worked for OMA on various large scale projects including RAK Gateway City – which won the 2009 Cityscape and International Business Award, BBC London headquarter strategic planning, and Interlace – a residential project in Singapore (*a+u* 15:09). As project architect for OMA's Beijing office, he worked on the tallest tower complex in Bangkok – MahaNakhon. Wang Shuo left OMA in 2009 to focus on the practice of META-Project with partner Zhang Jing and Max Fu. He has developed a series of urban research projects and is actively extending the idea into multiple dimensions of contemporary medium, including writing, video, web¬site, art installation and exhibition.

王硕，**META-工作室**创立合伙人，以北京为实践基地的建筑师，城市研究学者，策展人。毕业于清华大学，并在美国莱斯大学取得建筑学硕士学位。曾参与荷兰大都会建筑设计事务所（OMA）的一系列城市规划及建筑项目，包括RAK Gateway City——荣获2009全球城市景观与商业地产大奖，BBC伦敦总部战略规划，以及新加坡凯德The Interlace创新住宅项目（见《a+u》15:09），并作为项目建筑师设计曼谷最高楼Maha-Nakhon项目。2009年离开OMA，与合伙人张婧和Max Fu一同专注于META-工作室的实践探索。他发展了一系列城市研究计划，并积极地将其拓展到多元的当代媒介中，包括文字、影像、网络、装置艺术以及展览。

Zhang Ke, who completed Master's degree in Architecture at the Harvard Guraduate School of Design (1998) and Master's degree of Architecture and Urban Design at Tsinghua University, School of Architecture (1996), founded **ZAO/standardarchitecture** in 2001, a leading new generation design firm engaged in practices of planning, architecture, landscape, and product design.
Based on a wide range of realized buildings and landscapes in the past ten years, the studio's work has emerged as the most critical and realistic practice among the youngest generation of Chinese architects and designers. Although ZAO/standardarchitecture's works often take exceptionally provocative visual results, their buildings and landscapes are always rooted in the historic and cultural settings with a degree of intellectual debate.

张轲，1996年于清华大学取得建筑学学士和硕士学位，后留学美国，于1998年获得哈佛大学建筑学硕士学位。2001年创立**标准营造建筑事务所**，成为中国目前新一代设计团队代表之一，实践涵盖城市规划、建筑、景观、室内及产品设计等各种专业。
在过去10年的一系列重要文化及景观建成项目的基础上，标准营造以极具现实主义批判精神的实践作品在中国年轻一代建筑师中崭露头角。标准营造的建筑和景观作品往往具有独特视觉效果的同时，又极具思辨态度地根植于场所的历史文化背景。

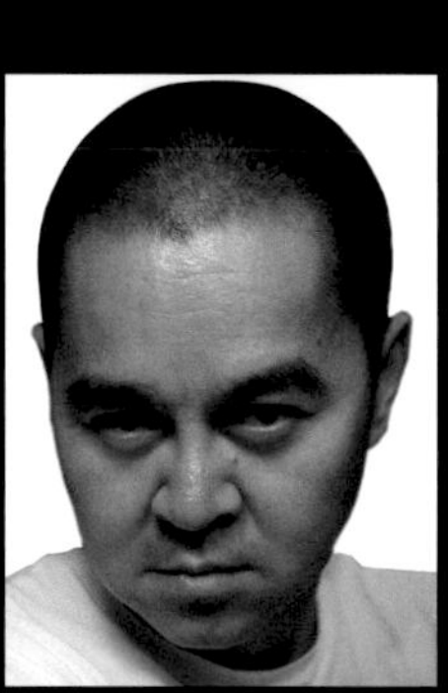

Chen Haoru, born in 1972 in Hangzhou, is currently the principal of Shanshang Jianzhu atelier. He graduated from New York University and China Academy of Art, and currently holds professorship in the School of Architecture of China Academy of Art. He is also the guest professor of Southeast University of Nanjing, guest design instructor of Rhode Island school of Design, and creative director of the Sun Farming Commune in Lin'an.

陈浩如，1972年出生于浙江省杭州市，山上建筑工作室主持建筑师，毕业于纽约大学和中国美术学院。现为中国美术学院建筑艺术学院教授，并兼任南京东南大学客座教授、美国罗德岛设计学院设计导师以及"临安太阳公社"创意总监。

Yu Ting is a registered Architect and holds a PhD in Architecture from Tongji University. He is a senior member of the Architectural Society of China. Amongst other commercial roles, Yu Ting sits as an expert panel member for Shanghai Construction Engineering. Yu Ting is heavily involved in science, research and technology and is a Professor of Engineering. His personal interests include gastronomy, writing, painting, history and fashion.
Yu Ting was awarded the supreme architectural prize by the Architectural Society of China as part of the 60th Anniversary Celebrations of the People's Republic of China. Whilst he has won many national and international prizes, Yu Ting is more interested in finding the correct balance between life and work, and to making a contribution to the built form of Shanghai using his skills and expertise.

俞挺，同济大学建筑学博士、教授级高工、国家一级注册建筑师、上海建设工程评标专家、中国建筑学会资深会员。广泛涉猎科学技术研究等领域，同时专注于美食、写作、绘画、历史和时尚产业研究等。
曾获得1949–2009建国60周年建筑创作大奖等多个国内外奖项，他在通过自身的专业知识技能致力于上海建筑业态发展的同时，也注重在日常生活与工作之间找寻恰好的平衡。

Li Xinggang is currently the chief architect of China Architecture Design Institute, the director of **Atelier Li Xinggang**, visiting professor of Tianjin University and the Southeast University, and the design tutor in School of Architecture Tsinghua University. He has won various honors and awards of architecture such as: China Youth Science and Technology Awards, International Architects Salon Awards, The Chicago Athenaeum International Architecture Awards. He was also invited to hold a mini-exhibition about his works, the "Geometry and Sheng Jing" (Studio-X Beijing, Architecture Centre of Columbia University, 2013), and to take part in some exhibitions, such as 11th Venice Biennale of Architecture (2008), "Illusion Into Reality: Chinese Gardens for Living" (2008), Dresden, "From Beijing to London: 16 Contemporary Chinese Architects" (2012), London, "Chinese Regional Architecture in a Post-Experimental Age" (2010, Karlsruhe/Prague).

李兴钢，中国建筑设计院总建筑师、**李兴钢建筑工作室**主持建筑师，天津大学、东南大学客座教授，清华大学建筑学院设计导师。曾获得中国青年科技奖、亚洲建筑推动奖、芝加哥雅典娜国际建筑奖；曾举办作品微展"胜景几何"（哥伦比亚大学北京建筑中心Studio-X，2013），并参加第11届威尼斯建筑双年展（2008）、德累斯顿"从幻象到现实：活的中国园林"展（2008）、伦敦"从北京到伦敦—当代中国建筑"展（2012）、卡尔斯鲁厄/布拉格"后实验时代的中国地域建筑"展（2010）等重要国际建筑及艺术展览。

Zhang Li (a.k.a. Li Brian Zhang / Brian Chang) (D. Eng., M. Arch.) was born in 1970. He is a Professor of Architecture and Chair of the Architecture Department in the School of Architecture, Tsinghua University, China. He leads the design practice Atelier **TeamMinus** in Beijing. He is currently a board member of the Architectural Society of China and the Editor-in-Chief of the leading Chinese magazine *World Architecture*. He is currently a board member of the Architectural Society of China and the Editor-in-Chief of the leading Chinese magazine World Architecture. He served as presenter of Venues, Beijing Winter Games 2022 Bidding Committee, 2015 Speaker.

张利，生于1970年，工学博士、建筑学硕士，清华大学建筑学院教授、建筑系副主任，**简盟工作室**主持建筑师，中国建筑学会理事，《世界建筑》的主编，同时担任北京申办2022年冬奥会的场馆规划与环境可持续性方面的负责人，先后在北京、洛桑和吉隆坡向国际奥委会进行了主题陈述。

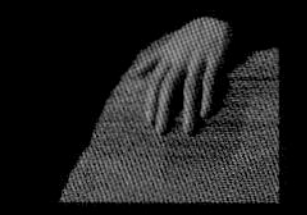

Zhao Yang was born in Chongqing, China, in 1980 and established Zhaoyang Studio in 2007 after graduating from Tshinghua University in Beijing in 2005. He received a master's degree from the Harvard Graduate School of Design in 2012 and founded **Zhaoyang Architects** in Dali, Yunnan, in the same year. Selected as the first Architecture Protégé of the Rolex Mentor and Protégé Arts Initiative in 2012 and designed the Home-for-All in Ohya for victims of the Great East Japan Earthquake under the direction of Kazuyo Sejima. Awarded the WA Chinese Architecture Award (2010). Invited to lecture at Tshinghua University, Tongji University, the Chinese University of Hong Kong, Art Basel, and Geoffrey Bawa memorial lecture, among other places. In 2015, invited to be one of the five participants for TOTO GALLERY MA 30th Anniversary Exhibition "The Asian Everyday: Possibilities in the Shifting World" in Tokyo.

赵扬，1980年出生于重庆市。2005年获清华大学建筑学硕士学位，2007年在北京创立**赵扬建筑工作室**。2012年毕业于哈佛大学，获建筑学硕士学位。同年，赵扬建筑工作室迁往云南大理。2010年获WA中国建筑奖优胜奖。2012年，赵扬作为首位获选"劳力士艺术导师计划"的建筑师，在著名建筑师妹岛和氏的指导下，完成了为在东日本大地震中受灾的灾民们而建的日本气仙沼市"共有之家"建筑项目。赵扬应邀在清华大学、同济大学、香港中文大学、巴塞尔艺术展和斯里兰卡杰弗里·巴瓦基金会等地举办学术讲座。2015年，赵扬建筑工作室应邀参加东京间画廊30周年纪念展"亚洲的日常——演变的世界的可能性"。

Yung Ho Chang, member of American Institute of Architects (AIA), is the founder, principal architect of **Atelier FCJZ**. Originally from Beijing, Chang received a Master of Architecture degree from the University of California at Berkeley in 1984. Became a licensed architect in the U.S. in 1989. He has been practicing in China since 1992 and established Atelier FCJZ in 1993. He has won a number of prizes, such as First Place in the Shinkenchiku Residential Design Competition in 1987, the Progressive Architecture Citation Award in 1996, and the 2000 UNESCO Prize for the Promotion of the Arts.

He has taught at various architecture schools in the US and China; he is presently a professor at Tongji University, and MIT. In 2009 he was bestowed the honorary membership by AIA Hong Kong. Since 2011, he became a Pritzker Prize Jury member.

张永和，美国建筑师学会会员（AIA），**非常建筑**创始人、主持建筑师。出生于北京，1984年在加州大学伯克利分校获得建筑学硕士学位。1989年在美国成为注册建筑师。1992年他在中国开始实践并于1993年创立非常建筑。荣获1987年日本《新建筑》国际设计竞赛一等奖，1996年度美国PA进步建筑奖和2000年UNESCO艺术贡献奖等系列奖项。

他曾执教于美国和中国的众多建筑院校，目前担任同济大学和麻省理工学院教授。2009年被美国建筑师学会香港分会授予荣誉会员。2011年起担任普利兹克建筑奖评委。

Wang Hao, co-founder of Rùn Atelier, was born in 1978. He graduated from Tongji University in 2002. He is a guest lecturer at School of Architecture of China Academy of Art. He won the Johans Göerder of Germany in 2004, and the German Bauwelt Prize for the First Work in 2013. He held personal exhibition "Free Structure - Chinese New Residents" at Shanghai Design Centre in 2013, and founded the Country Construction Institution with Zuo Jing (curator, publisher) in 2015, aiming to rebuild vernacular dwellings and create academies, while also encouraging vernacular housing researches and private timber structure design workshops.

王灏，生于1978年，润・建筑工作室的联合创始人，2002年毕业于同济大学，现执教于中国美术学院建筑艺术学院。曾获得2004年德国Johans Göerder 奖，2013年德国Bauwelt 处女作奖。2013年在上海设计中心举办“自由结构——中国新民居”个展；2015年与策展人以及出版人左靖一起成立了“乡村建造学社”，力图重造民居、营造书院，倡导民居研究以及木构设计的民间工作营。

Li Xiaodong, graduated from School of Architecture at Tsinghua University in 1984 and did his PhD at the School of Architecture, Delft University of Technology between 1989–1993, is a practicing architect, educator and researcher on architecture. Li Xiaodong's design ranges from interior, architecture to urban spaces. His works have received various awards including Bridge School (2008) in Fujian Province which won the 2009 Architecture Review Emerging Architecture Awards, 2010 Aga Khan Award for Architecture, Yuhu elementary school which won UESCO Jury Award for Innovation, ARASIA Gold Medal. Currently, he is now the chair professor of the architecture program at the school of Architecture, Tsinghua University, in Beijing. He is also a researcher, his publication in articles and books in both English and Chinese covers wide range of interests, from cultural studies, history and theory of architecture to urban studies.

李晓东，1984年毕业于清华大学建筑系，1989年至1993年在荷兰代尔夫特理工大学攻读建筑学博士学位。职业建筑师、建筑教育工作者与研究者。李晓东的设计涵盖室内设计、建筑设计和城市空间设计等。作品曾多次获奖，其中位于福建省的“桥上书屋”（2008）获得2009年英国“建筑回顾”世界新锐建筑奖一等奖和2010年阿卡汗建筑大奖，“玉湖完小”获得2005年联合国教科文组织亚太区文化遗产奖评审团创新奖、亚洲建筑金奖。目前，他担任北京清华大学建筑学院教授，同时专注于研究，发表了各类中英文文章并被收录在多本专业杂志与书籍内，其论文和研究涉猎建筑文化研究、建筑历史与理论、城市研究等方面。

Founded in 1999, under the leadership of partners Liu Xiaodu (top right), Meng Yan (top left) and Wang Hui, **Urbanus** is recognized as one of the most influential architecture practices in China and developed its branches in Beijing, Shenzhen. The design work of Urbanus ranges from cultural, educational, civic and residential projects and public spaces, cultural industry park to urban design projects, including large-scale urban complexes and renovations of existing urban areas. The design work of Urbanus have become new landmarks of urban life. The projects have received numerous prominent awards, and have been exhibited and published both nationally and internationally. The Tulou Collective Housing project was designed under the leadership of Liu Xiaodu and Meng Yan.

都市实践，当今中国最具影响力的建筑师团队之一，由刘晓都（右上）、孟岩（左上）和王辉创建于1999年，目前在北京、深圳设有分支机构。都市实践的设计项目涵盖博物馆、学校、住宅、公共空间、政府办公、大型城市综合体、文化产业园区、旧城改造等多个类别。多项建成作品成为城市生活新地标，获得重要建筑奖项，并在世界各地参展及出版。社会住宅“土楼公舍”由合伙人刘晓都和孟岩共同主持完成。

Zhu Jingxiang, associate professor in School of Architecture, the Chinese University of Hong Kong (CUHK), received his education in Southeast university and the Swiss Federal Institute of Technology (ETH) Zurich. Before joined CUHK in 2004, he taught in Southeast University and Nanjing University, worked as an architect in mainland for 10 years. His early award-winning public building works were widely exhibited.

His research specialty is in the area of new articulation of structures and space, light weight building system, cost-effective architecture and vernacular construction. Since 2008, he invented various innovative prefab light weight building systems, and applied them in quake areas and developing areas. These works help him win several influential Chinese academic awards, 2012 China Innovator of the Year Award in architecture from *The Wall Street Journal*, and 2015 Hong Kong Construction Industry Council Innovation award.

朱竞翔，香港中文大学建筑学院副教授。先后就读于东南大学建筑系和苏黎世瑞士联邦理工学院，曾执教于东南大学与南京大学，并在内地有10年的建筑师从业经验。早期的公共建筑获奖作品曾被广泛展出。

目前专注于研究新型的空间结构、轻型建筑系统和可持续建筑技术。2008年以来，他的团队发展了多种新型的预制轻型建筑系统，包括学校、保护工作站及办公建筑项目，并成功应用在地震区和发展中国家地区。这些工作使他获得诸多有影响力的中国学术奖项，包括2012年《华尔街日报》中文版“中国创新人物奖”、2015年香港建造业议会创新奖一等奖等。

OPEN Architecture was founded by Li Hu (left) and Huang Wenjing in New York City. It established the Beijing office in 2006. OPEN Architecture was recognized by many architectural awards for its progressing work, including *World Architecture*'s Chinese Architecture Award 2012 and 2014, Asia Pacific Interior Design Awards 2013 and most recently, Merit Award of 2015 AIANY Design Awards.
Li Hu received his B. Arch. from Tsinghua University in Beijing in 1996 and M. Arch. from Rice University in 1998. He is former director of Columbia University GSAPP's Studio-X Beijing, and former partner of Steven Holl Architects. Huang Wenjing received B. Arch. from Tsinghua University in Beijing in 1996, and M. Arch. from Princeton University in 1999. She is a registered architect of New York State and a member of AIA.

OPEN建筑事务所由李虎（左）和黄文菁创立于纽约，2006年建立北京工作室。OPEN的作品获得2012年和2014年WA中国建筑奖优胜奖、2013年亚太区室内设计大奖（APIDA）公共空间类金奖以及美国建筑师协会纽约分会2015年优秀设计奖，并入围伦敦设计博物馆2015年度设计奖。
李虎，1996年取得清华大学建筑学学士学位，1998年取得美国莱斯大学建筑学硕士学位。曾任Studio-X哥伦比亚大学北京建筑中心负责人以及美国斯蒂文·霍尔建筑事务所合伙人。黄文菁，1996年取得清华大学建筑学学士学位，1999年取得美国普林斯顿大学建筑学硕士学位。她是美国纽约州注册建筑师、美国建筑师协会会员。

Tao Lei, who graduated from China Central Academy of Fine arts in 2002, founded **Tao Lei Architecture Design** (TAOA) in 2007. He has been constantly seeking a new way of architecture which not only coexists with natural environment but could adapt itself to the future. Based on the combination of traditional Chinese culture and modern notion, the value of his projects is to create free and authentic space atmosphere in an organic and relaxing way. Such value enables the team to perfect every single segment of architectural practice involving project planning, conceptual design and construction detail. Furthermore, TAOA pays attention to the social context of architecture and meets clients' demand accordingly.

陶磊，2002年毕业于中央美术学院建筑学院，2007年成立**陶磊建筑事务所**（TAOA）。在实践中，陶磊不断寻求一条与当今社会及自然环境共存的道路，并以东方的自然观探讨属于未来的建筑。在中国传统的空间意识、文化意识及当下价值观的前提下，其项目能以有机、轻松且快乐的方式体现真实、自由的生活氛围，营造属于自己的空间意境。无论是居住、商业还是文化中心，TAOA重视从项目策划、概念设计到项目施工过程中的每一个环节，一直致力于对建筑品质的完美追求，重视建筑的工程细节和最终真实建筑的完成度，并注重业主的真实需求与建筑表达社会意义的统一关系。

Natural Build Operation LLC, founded by Shui yanfei (top left), Ma Yuanrong (top right) and Su Yi-Chi in 2011, is a design practice based in Shanghai that operates in the fields of architecture and beyond. The office is founded on a commitment to the naturalness of the built environment, and the transformative potential of nature on architecture. Without submitting to any arbitrary systematic or aesthetic convention, the work seeks a naturality in design and build through the use of analytical and constructive techniques.

直造建筑事务所由水雁飞（左上）、马圆融（右上）、苏亦奇于2011年在纽约成立，现立足于上海，涉及建筑及其相关领域。事务所致力于人造环境的自然化，并尊重自然作用于建筑变化的可能性。不从属于任何先验的系统或美学惯例，通过理性推演及物化研究的工作方式，探寻恢复设计与建造的自然性。

Tong Ming is the principle architect of a Shanghai based architectural office, **TM Studio**, projects of which include museums, residences, institutional and commercial projects as well as urban and community specific development plans and studies. His works has been published widely and participated exhibitions include: Contemporary China, Rotterdam (2006); Biennale of Hong Kong (2007); M8 in China,German Architecture Museum, Frankfurt. "Dans la ville chinoise", Paris (2008), Venice Biennale (2008, 2014); Triennale of Milan (2012). Meanwhile, Tong Ming is also now an urban design professor in the College of Architecture and Urban Planning Tongji University, responsible for teaching and research in urban design, and urban regeneration.

童明，**童明工作室**（TMStudio）的主持建筑师。TMStudio以上海为实践基地，其作品涉及文化、居住、办公、商业以及城市设计等诸多领域。TMStudio的建筑作品曾得到广泛发表并参加过许多国内外的重要展览，其中包括2006年鹿特丹当代中国建筑展，2007年香港双年展，2009年法兰克福建筑博物馆的“M8在中国”展，2008年和2014年威尼斯双年展，2012年米兰三年展。童明同时担任同济大学建筑与城市规划学院教授，负责城市设计、城市更新等领域的研究与教学工作。

Zhu Xiaofeng received his Bachelor's degree in Architecture from Shenzhen University and Master's degree of architecture from the Harvard Graduate School of Design. He founded **Scenic Architecture Office** in Shanghai in 2004. He has been teaching at Tongji University as visiting professor since 2012.
Scenic Architecture office believes that the spirit of architecture exists in how people perceive the basics of nature and living. Architecture of 21st century shall not only respond to human's needs, but also act as a positive media between human and environment. They use architecture to explore how space and time stimulate and absorb each other, and how to establish balanced and dynamic relevance among human, nature and society. With persistent efforts on practice and thoughts, we are on the way to make architecture as a yearning life experience, and a carrier of both material and spirit.

祝晓峰，深圳大学建筑学学士，哈佛大学建筑学硕士。2004年在上海创办**山水秀建筑事务所**。2012年起担任同济大学客座教授。
山水秀认为建筑来源于人对自然和生活最基本的感知。21世纪的建筑不仅要响应人的需求，更要积极担当人与环境之间的媒介。山水秀用建筑语言寻求空间与时间的相生、相融，寻求在人、自然及社会之间建立平衡而又充满生机的关联，通过思想的进步和实践的积累，使建筑成为令人向往的生活体验，成为物质和精神的共同载体。

Atelier Archmixing, based in Shanghai, was founded by Zhuang Shen (top left) and Ren Hao (bottom left) in 2009, joined Tang Yu (bottom right) and Zhu Jie as partners. They build their reputation on innovative and flexible design strategies deeply rooted in China's complicated urban and rural context. Their work boasts a broad range of typologies and scales, and has been well awarded and extensively exhibited and published, both domestically and abroad. The principal architect Zhuang Shen is recognized as one of the pioneering young Chinese architects. He is now also a guest professor in College of Architecture and Urban Planning, Tongji University.

阿科米星建筑设计事务所于2009年由庄慎（左上）、任皓（左下）在上海创立，唐煜（右下）、朱捷为合伙人。他们创新而灵活的实践策略根植于中国复杂的城乡背景中，并因此建立了自己的声誉。
阿科米星的设计无论是内容还是范围都不拘一格，受到国内外重要专业媒体的持续关注与广泛好评。主持建筑师庄慎是一位具有开创性的本土年轻设计师，自2014年起兼任同济大学建筑与城市规划学院客座教授。

Atelier Deshaus

Long Museum West Bund

Xuhui District, Shanghai, China 2013

大舍建筑设计事务所

龙美术馆西岸馆

中国，上海市，徐汇区 2013

Feature:
Architects in China

Atelier Deshaus
Long Museum West Bund
Xuhui District, Shanghai, China

pp. 20–21: Main entrance of the museum on the northwest side. The coal hopper unloading bridge built in 1950s is preserved. Photo by Xia Zhi. This page: Southwest facade. Opposite: The red footbridge connect the courtyard on the second floor to the terrace on the unloading bridge. Two photos by Su Shengliang. All photos on pp. 20–27 courtesy by the architects.

20–21页：美术馆西北侧的主入口。建筑保留了建于20世纪50年代的煤漏斗卸载桥。本页：西南立面。对页：红色的天桥连接了二楼的庭院和卸载桥的露台。

Credits and Data

Project title: Long Museum West Bund
Program: Art museum, new project with partial reconstruction
Location: No. 3398, Longteng Avenue, Xuhui District, Shanghai, China
Completion: December 2013
Architects: Atelier Deshaus
Principal architect: Liu Yichun, Chen Yifeng
Construction area: 33,007 m²

This page: Interior view of the contemporary art gallery on the ground floor from the south. Vaults with different directions and heights achieve the diversity of spatial character. Photo by Su Shengliang. Opposite: View of the exterior space under the vaults next to the entrance from the north. The stairs on the left connect to underground. Photo by Xia Zhi.

本页：从南侧看首层当代艺术展厅的内部。拥有不同高度和方向的拱顶丰富了空间的特征。对页：从北侧看入口处拱顶下方的室外空间。左边的楼梯连接了地下空间。

龙艺术商店

1. Museum entrance
2. Entrance hall
3. Shop
4. Contemporary art gallery
5. Video room
6. Cloakroom
7. Service room
8. Void
9. Temporary gallery
10. Art and design shop
11. Resteraunt
12. Vip room
13. Freight elevator
14. Coal hopper unloading bridge space
15. Art work by Xu Zhen: Sports Field

Ground floor plan (scale: 1/2,000) / 首层平面图（比例：1/2,000）

1. Existing basement B2
2. Existiong floor slab B1
3. Existing main beam ground floor
4. New floor slab ground floor
5. New as-cast-finish concrete wall
6. Underfloor air supply system air outlet
7. Air conditioning inlet interface
8. Large-space intelligent active control sprinkler

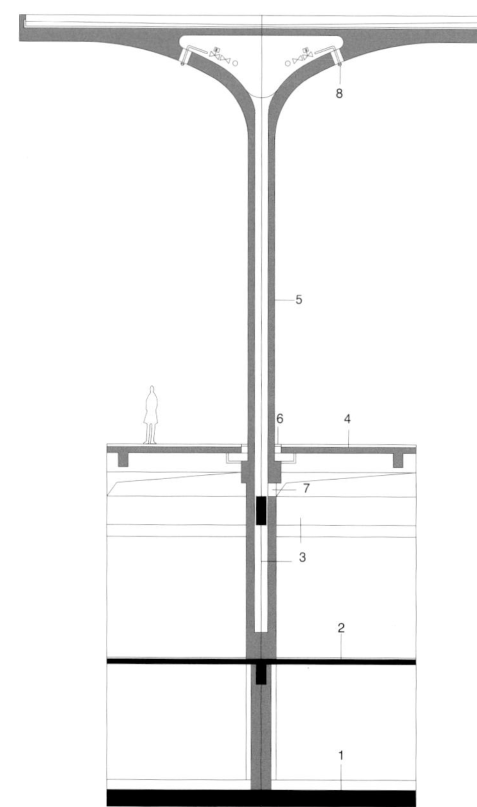

Partial section detail (scale: 1/250) / 部分细节剖面图（比例：1/250）

Opposite: The contemporary art gallery on the second floor. This page: View of the red terrace floating on the unloading bridge from the southeast. Two photos by Su Shengliang.

对页：二楼的当代艺术展厅。本页：从东南侧看悬浮在卸载桥之上的红色露台。

Long Museum West Bund is located on the banks of the Huangpu River, Xuhui District, Shanghai Municipality, on a site which was used as a wharf for coal transportation. Before the commencement of the design, there remained a coal hopper unloading bridge of about 110 m in length, 10 m in width and 8 m in height, which was constructed in the 1950s, with a two-story underground parking area completed in 2013.

The new design adopts a cantilever structure featuring a "vault-umbrella" with independent walls, while the shear walls with free layout are embedded into the original basement so as to be built with the original framework structure. With the shear walls, the first underground floor of the original parking has been transformed into an exhibition space, with the aboveground space highlighting multiple orientations because of the relative connection of the "vault-umbrella" at different directions. Also, the electrical and mechanical system has been integrated in the "vault-umbrella" structure. For the aboveground space covered by the "vault-umbrella", the walls and ceiling are as-cast-finish concrete surface so that their geometrical dividing line seems faint. Such a structure not only shields the human body but also visually echoes the coal hopper unloading bridge at the wharf. Moreover, the building's internal space can also represent a kind of primordial and timeless charm while the spatial dimension, large or small, and the as-cast-finish concrete surface with the seam among molding boards and the bolt holes bring a sense of reality as well. The directness and simplicity resulting from this "literal" structure, material and space plus the sense of force or lightness because of the large-scale overhanging style enables the overall building's continuation of the industrial property of the original site, not only in time but in space.

The flowing exhibition space under the aboveground as-cast-finish concrete "vault-umbrella" and the "white box" exhibition space on the first underground floor are connected with spiral ladders downward. The parallel tensility highlighting the space, primordial but realistic, and the art exhibitions from the ancient, modern and contemporary periods, display an exhibition space featuring temporality.

龙美术馆西岸馆位于上海市徐汇区的黄浦江滨，基地以前是运煤的码头。设计开始时，基地现场有一列被保留的建于 20 世纪 50 年代的煤漏斗卸载桥（大约长 110m、宽 10m、高 8m）和一个完成于 2013 年的两层地下停车库。

新设计采用了独立墙体的"伞拱"悬挑结构，呈自由状布局的剪力墙插入原有的地下室，与原有框架结构柱浇筑在一起。原地下一层的车库空间由于剪力墙体的介入转换为展览空间，地面以上的空间由于"伞拱"在不同方向的相对连接，形成了多重的意义指向。机电系统都被整合在"伞拱"结构的空腔里，地面以上的"伞拱"覆盖空间、墙体和天花板的表面均为清水混凝土，这模糊了它们的几何分界位置。这样的结构性空间，在形态上，对人的身体形成庇护，亦与被保留的江边码头的煤漏斗卸载桥产生视觉呼应。建筑的内部空间也得以呈现出一种原始的野性魅力，而有着大小调节考量的空间尺度以及留有模板拼缝和螺栓孔的清水混凝土表面又会带来一种现实感。这种"直白"的结构、材料、空间所形成的直接性与朴素性，加之大尺度的出挑所产生的力量感和轻盈感，使整个建筑与原有场地的工业特质之间生成了一种时间与空间的接续关系。

地面以上的清水混凝土"伞拱"下的流动展览空间与地下一层传统"白盒子"式的展览空间，由一个呈螺旋回转、层层跌落的阶梯空间连接。既原始又现实的空间与古代、近代、现代直到当代艺术的展览陈列所内含的这种并置的张力，营造出了一种具有时间性的展览空间。

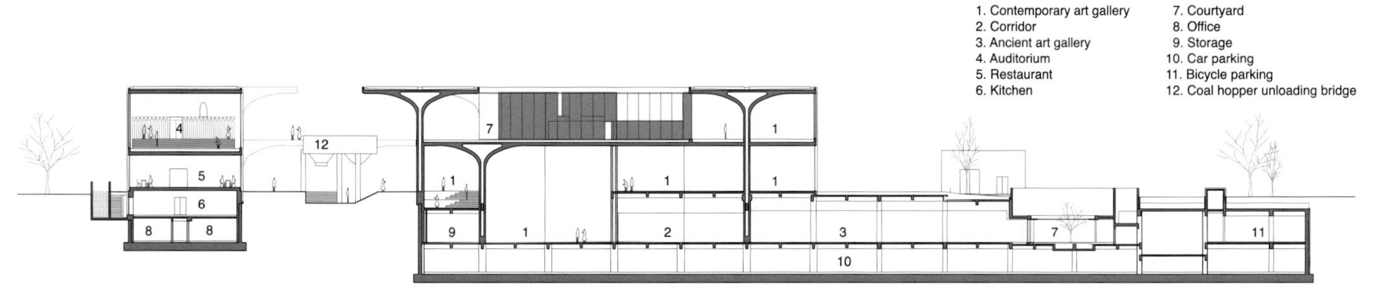

Longitudinal section (scale: 1/1,000) / 纵向剖面图（比例：1/1,000）

Trace Architecture Office (TAO)

Forest Building

Tongzhou District, Beijing, China 2014

迹・建筑事务所

林建筑

中国，北京市，通州区　2014

Trace Architecture Office (TAO)
Forest Building
Tongzhou District, Beijing, China

Located in a park full of trees by the Grand Canal River on the east side of Beijing, the project started without a clearly defined program. The client envisioned a flexible space for different functions (restaurant, bars, event space, gallery, office, etc.). Inspired by No-Stop City, a utopian urban design proposed by Archizoom in 1960s based on indeterminism, Forest Building explores a spatial system generated from additive units responding to the uncertainty of the program, as well as structural form and construction system responding to the character of the site.

As the program doesn't give much to initiate the design, the architects examined the site to find a starting point to capture the sense of place. The river and the trees were the strongest characteristics of the site, so the idea started to emerge: to create a place where people can sit under the tree and enjoy the view of the river – just like what everyone does in a park, even without architecture. This became the beginning point. Consequently, a base unit was developed. By organizing it in a grid system, the overall form was generated. Later on, the structure, material, landscape, mechanical system, drainage, all these elements were conceived and developed following the original intention.

The structure is made from the repetition of a basic unit, a tree-like column with four cantilevered beams at varied heights. The plan is based on a grid system that is flexible enough to adapt to the site and existing trees. Such flexibility also makes it possible to break down construction into a number of phases.

The structure is elevated by a floating concrete platform, which protects the glulam structure from moisture. At the same time, the utilities run underneath the concrete floor so that the wood ceiling is free to express the structure and space purely.

The architecture is mainly constructed from glue laminated wood and rammed earth. These natural materials can adjust the relative humidity and temperature of indoor and outdoor areas, and will biodegrade in the future. Clear glass windows fill in the space between wood columns and rammed earth walls, inviting people inside to experience the manmade forest and the natural forest outside at the same time.

林建筑项目位于北京东部的通州运河森林公园内。在设计的初始阶段，项目的功能定位并不明确，甲方将其憧憬为一个可同时适应各种不同功能的灵活空间（如具有餐厅、酒吧、会议活动场地、画廊和办公等功能）。受 20 世纪 60 年代 Archizoom 基于非决定论提出的乌托邦城市设想——No-Stop City 的启发，林建筑探索了一种基于单元的重复组合形成整体的空间体系、结构形式和建造体系，以回应场地特征，并适应项目在功能上的不确定性。

由于任务书未能带来明确的启示，建筑师便从场地中去寻找设计的出发点。对这片场地而言，运河与树林无疑是其最突出的特征。于是，一个想法自然而然地浮现出来：创造一个让人们就像处在公园里那样（即使公园里没有什么建筑），可坐在树下欣赏风景的场所，这即是该设计的出发点。之后是基础单元的设计，接着通过网格系统的组织，将基础单元进行组合，从而生成建筑的整体形态。最后，结构、材料、景观、设备系统和排水组织等所有要素均围绕原始的设计理念而逐步形成并发展起来。

以柱子为中心并伸出 4 条高度不等的悬臂梁的树型结构是林建筑的基本单元，它在合理的高度变化控制下重复组合成建筑的整体结构。平面是基于有些曲折的规律格网设计，它的边界自由，便于绕开场地现存的树木并可获得分期施工的灵活性。

建筑被设置在“飘浮”于地面的混凝土平台上，这样一方面有利于木结构防潮；另一方面，将机电设备服务层布置在平台之下，使屋顶下部解放出来，还原为纯粹的结构和空间。

建筑主体的材料为木和夯土，它们的自然质感呼应了场地中的树木和泥土。这些材料可以自然地呼吸，有效地调节室内外的相对湿度和温度，将来也可自然降解，回归自然。立面上，木和夯土墙之间以玻璃幕墙填充，为室内使用者提供同时身处人工与自然森林间的感受。

pp. 28–29: General view from the northeast. This project is located in a park full of trees by the Grand Canal River. Photo by Su Shengliang. pp. 30–31: Interior view from bar by the main entrance. A continuous view is created from the repetition of a basic unit in both the interior and exterior space. Photo by Su Shengliang. All photos on pp. 28–33 courtesy of the architect.

28–29页：从东北侧看到的全景。该项目坐落在大运河边绿树成荫的公园里。30–31页：从主入口的酒吧处看内部空间。通过基础模块在室内外的不断重复，营造出一种连贯的视觉体验。

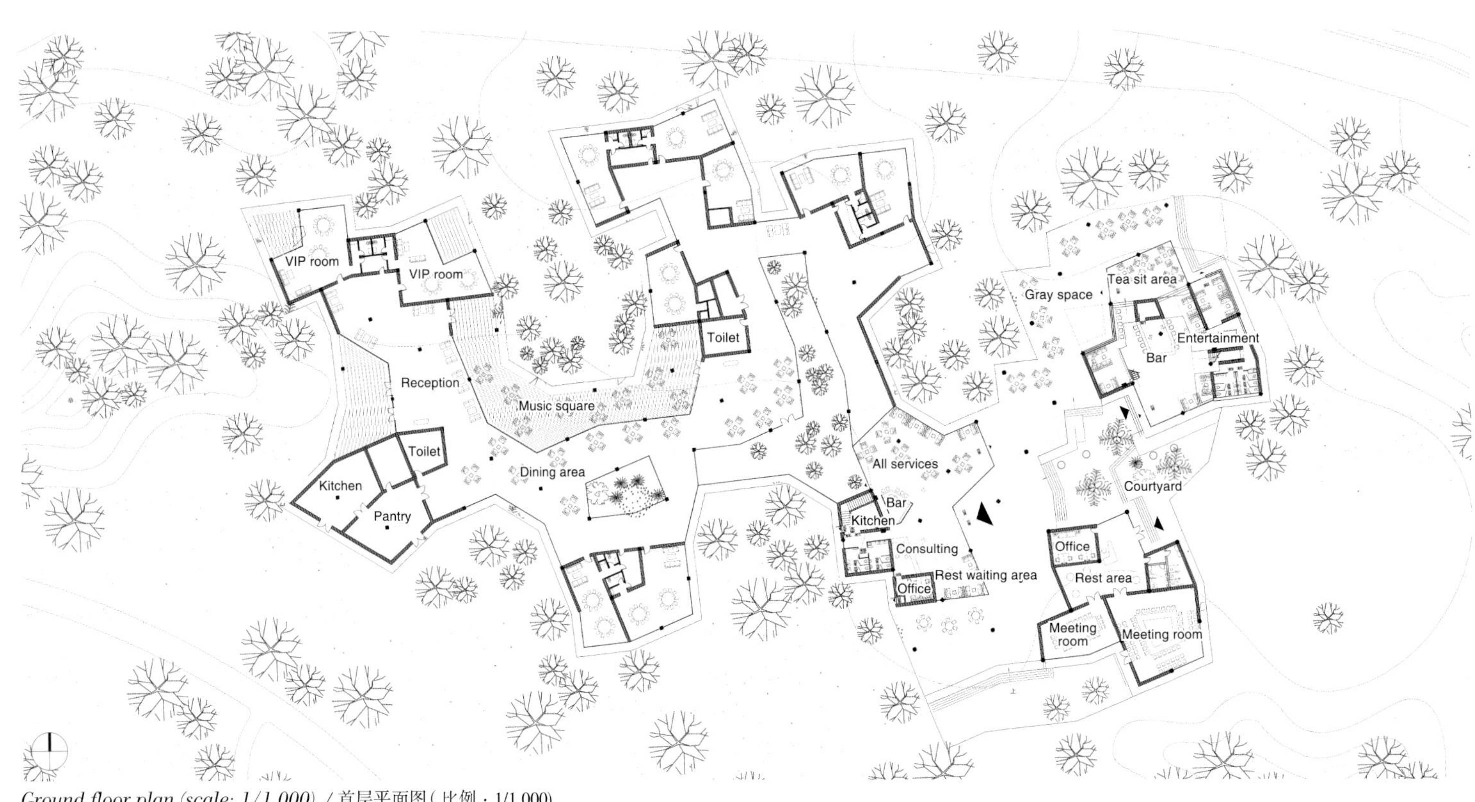

Ground floor plan (scale: 1/1,000) / 首层平面图（比例：1/1,000）

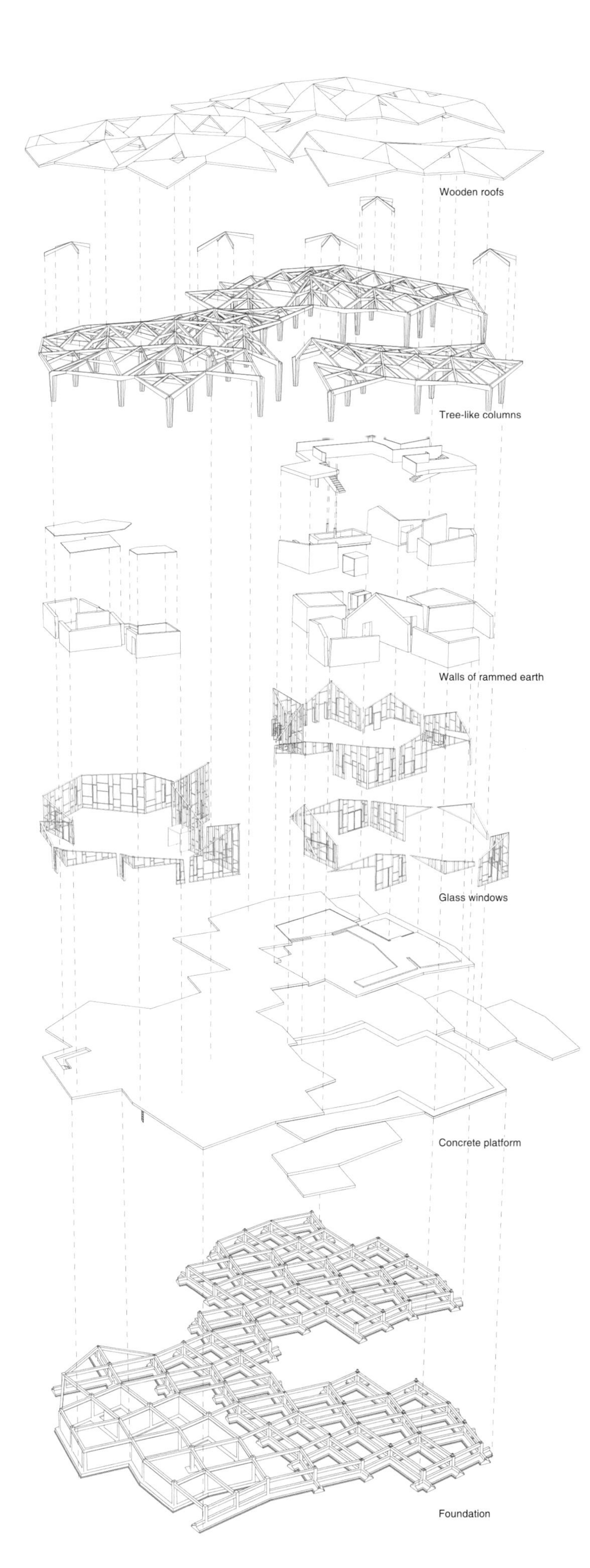

Extended axonometric drawing / 展开的轴测图

Credits and Data
Project title: Forest Building
Client: Beijing Meijingtiancheng Investment Co., Ltd.
Program: Reception, restaurant, conference, bar, office
Location: Grand Canal Forest Park, Tongzhou District, Beijing, China
Design: 2011–2013
Construction: 2012–2014
Architects: Trace Architecture Office (TAO)
Principal architect: Hua Li
Project team: Hua Li, Zhao Gang, Jiang Nan, Lai Erxun, Chen Kai, Alienor Zaffalon, Zhang Zhiming, Elisabet Aguilar Palau, Jodie Zhang, Bai Ting
Floor area: 4,000 m² (phase 1: 1,830 m²)
Structural system: Timber structure

This page, above: View of continued roofs. There are three variations in the height of columns: 3.35 m, 4.35 m and 5.35 m. Photo by Xia Zhi. This page, below: Study model of column, 1/10 scale.

本页，上：连续的屋顶。柱子有3种不同的高度：3.35m，4.35m和5.35m。本页，下：柱子的研究模型，比例为1/10。

Zhang Lei

Nanjing Wanjing Garden Chapel

Nanjing, Jiangsu, China 2014

张雷

南京万景园小教堂

中国，江苏省，南京市 2014

This page: View of the lake from the north cornner of the corridor in the chapel. It is located in the Nanjing Wanjing Garden Chapel. Opposite: North facade from the lake. All photos on pp. 34–39 by Yao Li, courtesy of the architect.

本页：从小教堂走廊的北角看向湖面。教堂坐落在南京万景园中。对页：从湖面看北侧立面。

Zhang Lei
Nanjing Wanjing Garden Chapel
Nanjing, Jiangsu, China

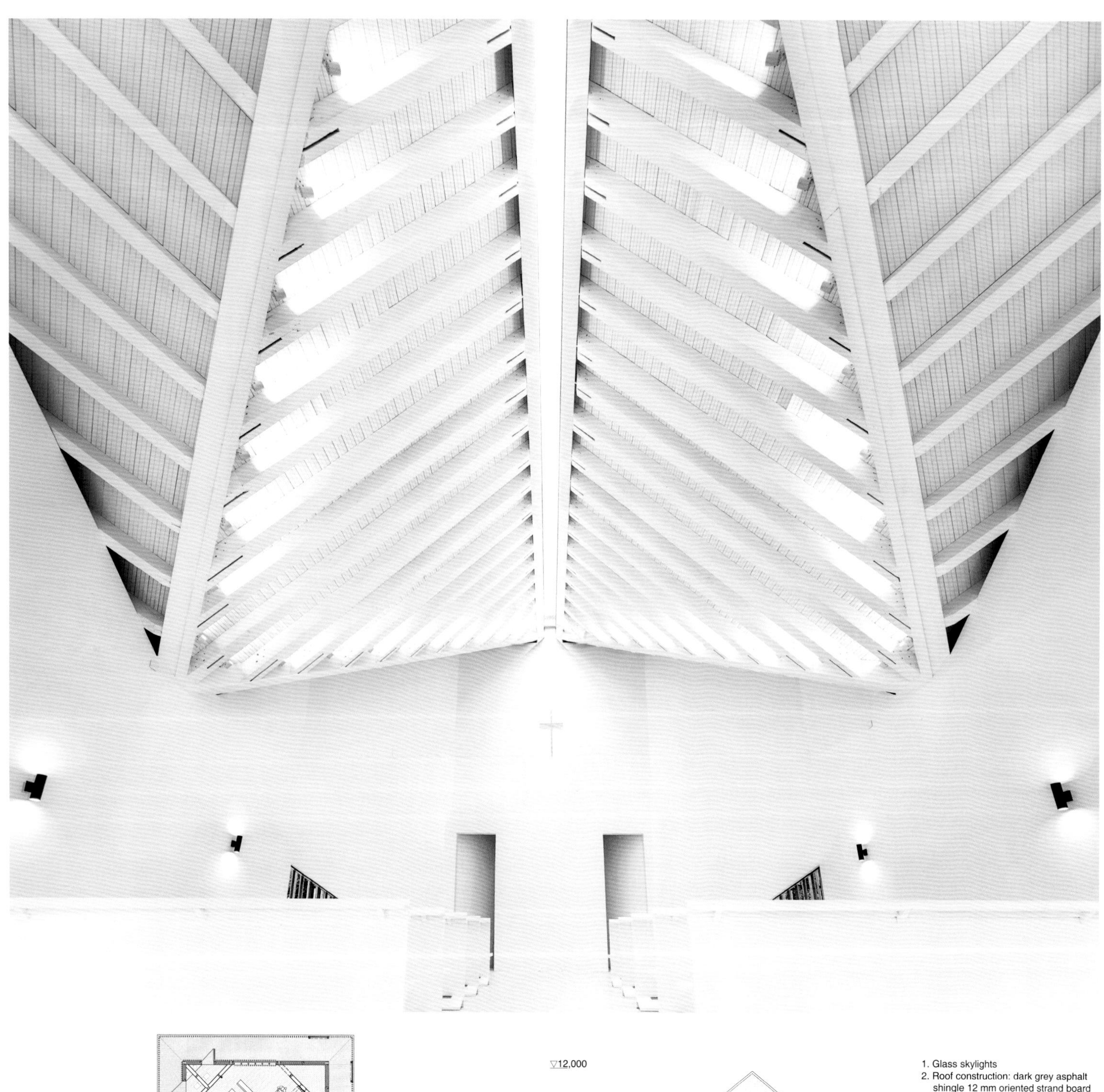

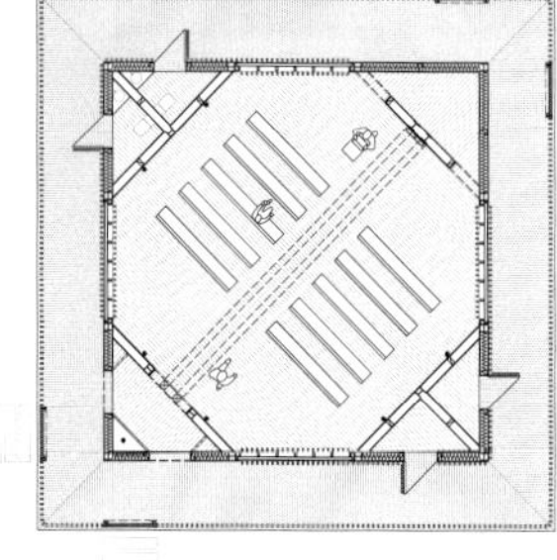

Ground floor plan (scale: 1/400) / 首层平面图（比例：1/400）

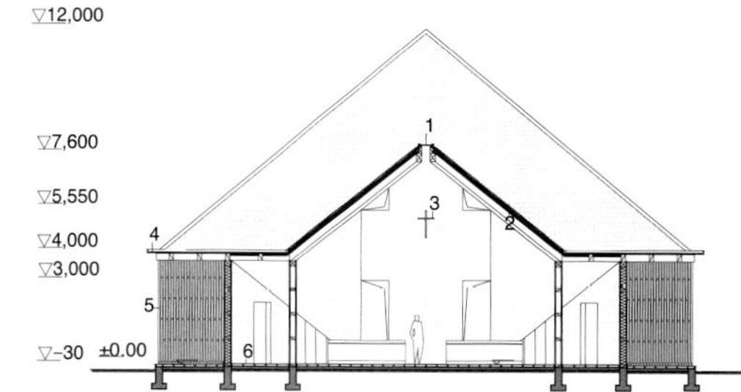

1. Glass skylights
2. Roof construction: dark grey asphalt shingle 12 mm oriented strand board 38 / 89 mm wood keel 15 mm oriented strand board wood truss
3. 38 / 89 mm SPF wooden cross
4. Galvanized roof gutter
5. 38 / 89 mm SPF wooden grid 6 ground construction:10 mm anticorrosive wood floor 50 / 100 mm wood keel 40 mm fine stone concrete 150 mm gravel compaction soil compaction

Section (scale: 1/400) / 剖面图（比例：1/400）

The project, a 200-m² small chapel, is located in Wanjing Garden along Nanjing's Riverfront. Hosted by priests from Nanjing Union Theological Seminary, it supports religious activities like worship and wedding services. This wood and steel structured chapel has a gentle exterior shape as well as strong interior space infused with mysterious religious power. Its plain material doesn't fail in expressing the delicate construction logic.

The earliest and fundamental church space has two interrelated tendencies – centre and depth. Both the centrality of the Pantheon and the deep axis of the Basilica were inherited in early Christian architecture. The central and axial symmetry of church space became less prominent during the Modernist period, which was closely related to the Protestants' rejection of Catholic hierarchy. In the design of Nanjing Wanjing Garden Chapel, the architect didn't purposefully reject the "centrality" and "depth" of the classical space sequence. In plan, it has an octagonal central hall surrounded by square shaped corridors. In section, the use of the V-shape profile of the roof and the long slit of the skylight amplify the depth of space and emphasize the high point over the sacred space where all axes meet.

The wooden corridor forms a unique double shell of the central hall space. The inner shell is more closed, allowing natural light to penetrate only through openings on roof and walls. The outer shell, composed of delicate SPF strips, serves as a filter of the view outside, implying the start of a religious spatial experience. The double shell system creates a uniquely interesting oriental architectural space that's different from the "closedness" of traditional stone chapel, or the "openness" of modern architecture. The small chapel has a perfectly square-shaped plan. Even with the 45-degree turning connecting inner space to exterior structure, the square remains highly integrated, symmetrical and centre oriented. The hidden diagonal axis in the plan also extends to the roof structure, realizing a complete integrity of space, material and power.

A "light" construction strategy is a wise choice under the tight project schedule and limited budget. The roof structure adds expressiveness to the entire space. All interior surfaces are painted white, emphasizing the leading role of light and space. The exterior wood strips and asphalt shingles are left in their natural color, emphasizing the importance of nature.

As a religious institution with very simple functions, this small chapel has its overly "ideal" space that couldn't attribute to any particular style, but "the nature that tolerates all."

1. Eaves construction: dark grey aluminum veneer sealing side dark grey steel fittings, the retaining bolt
2. Roof construction: double agate black asphalt shingle roof, 3 mm self-adhesive waterproofing materials, 11.9 mm OSB, 38 / 89 mm wood keel @610, 15 mm J SPF wood, wood truss
3. 38 / 89 mm SPF wooden grid
4. Roof construction: double agate black asphalt shingle roof, 3 mm self-adhesive waterproofing materials, 11.9 mm OSB, 38 / 89 mm wood keel @ 610 (embedded glass wool),15 mm J SPF wood, wood truss
5. Wall construction: texture coating, 10 mm cement pressure plate, 25 / 38 mm anticorrosion wooden keel (the vertical laying) @ 406, vapour-permeable membrane, 9.5 mm OSB, 38 / 140 mm SPF wooden grid (embedded 139 mineral-wool insulation), single 15 mm fire prevention plasterboard, interior wall coating
6. 38 / 49 mm SPF wooden grid
7. SPF fittings
8. Ground construction:10 mm anticorrosive wood floor, 40 / 80 mm wooden keel, 40 mm C20 fine stone concrete, 10 mm 1:3 cement mortar, 1.5 mm waterproof membrane, 60 mm C15 fine stone concrete cushion layer, pave durable plastic film layer, 150 mm gravel, soil compaction

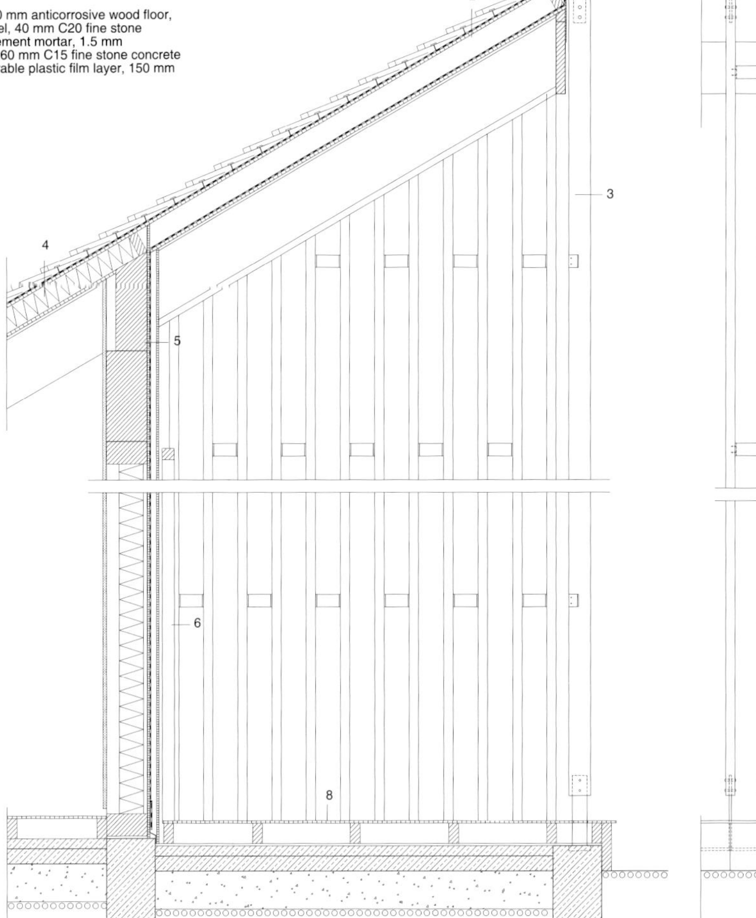

Section detail (scale: 1/25) / 剖面细节图（比例：1/25）

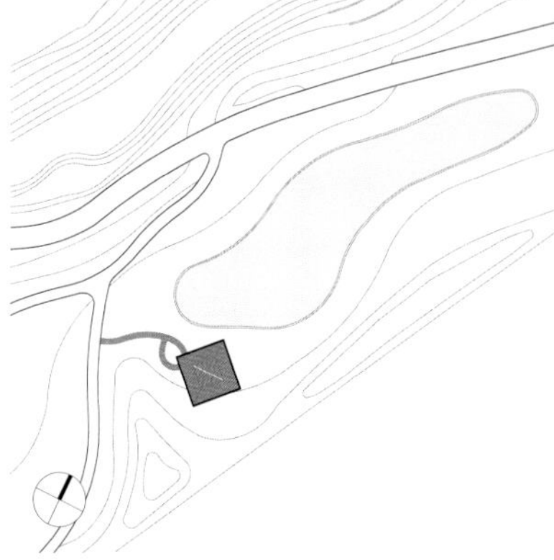

Site plan (scale: 1/3,000)
总平面图（比例：1/3,000）

p. 36: Interior view of the chapel from the entrance to the altar. Natural light enters from the top light between 300 mm × 200 mm steel beams. p. 37: Interior view towards the north corner of the chapel. Opposite: Evenig view from the west.

36页：从入口处面向祭坛看教堂内部。自然光线穿过钢梁之间300mm × 200mm的天窗。37页：看向教堂内部北侧。对页：从西侧看的夜景。

项目位于南京滨江风光带的万景园内，是一个面积仅为 200 m² 的小教堂。该教堂由南京金陵协和神学院的牧师主持，满足了信众聚会、举办婚礼等功能需求。这个钢木结构的小教堂具有平和的外形和充满神秘宗教力量的内部空间，质朴的材料和精致的空间逻辑构造，诠释了建筑师一贯的建筑观——对立统一。

最早也是最基本的教堂空间布局存在两种相互关联的倾向：“集中”和“纵深”。源自万神殿的集中性和巴西利卡的纵深的空间序列，都在早期基督教建筑中得以继承和延续。现代主义时期之后，明确中心和轴线对称的教堂空间组织形式变得不那么突出，这与新教各派拒绝天主教的教阶体制、崇尚简朴不无关系。在万景园小教堂的设计中，建筑师并未有意排斥“集中”和“纵深”的古典空间序列。小教堂的平面是强调集中性的正方形回廊和正八边形的主厅，而剖面由于折板屋顶的限定，以及南北向屋脊中央的狭长天窗的光带，显示出强烈的纵深空间感，并且突出了圣坛上方空间向上高耸的轴线焦点。

小教堂设计独特的回廊空间形成了主厅空间的双层外壳。内壳封闭，突出了来自顶部和圣坛墙面裂缝的纯净的天光效果；外壳因其精密的 SPF 格栅构造，成为外部风景的过滤器和内部宗教场所体验开始的暗示。双层外壳的空间边界，不同于传统石质教堂的“内向”和经典现代建筑的“外向”，它带有独特的东方建筑空间趣味。小教堂首先具有一个完美的正方形平面。内部空间和外部结构之间虽存在 45°的转角，但这个正方形平面仍保持了高度的完整性、对称性和向心性。暗藏的对角线延伸到屋顶结构，产生了精致的折板屋面，它同样是空间、力和材料的高度统一。

“轻”的建造策略是建筑师在工期紧张和造价有限的条件下作出的明智选择。屋顶的钢木结构赋予了空间丰富的表现力。内部所有的表面都涂饰白色，将主角让与空间和光。外部的所有材料——木质格栅、沥青瓦屋面，均保持原色并等待时间的印记，将主角让与大自然。

作为一个功能简单的日常宗教活动场所，小教堂的空间过于“理想”而无法被解释为某种特定的宗派。建筑师之所以能为其展开有效的设计，也许是因为信奉了“包容一切的自然”。

Credits and Data

Project title: Nanjing Wanjing Garden Chapel
Location: Nanjing, Jiangsu, China
Project year: 2014
Architect: Zhang Lei
Collaborator: Architectural Design & planning Institute, NJU
Project area: 200 m²

Vector Architects

Seashore Library

Beidaihe New District, Qinhuangdao, Hebei, China 2015

直向建筑事务所

三联海边图书馆

中国，河北省，秦皇岛市，北戴河新区 2015

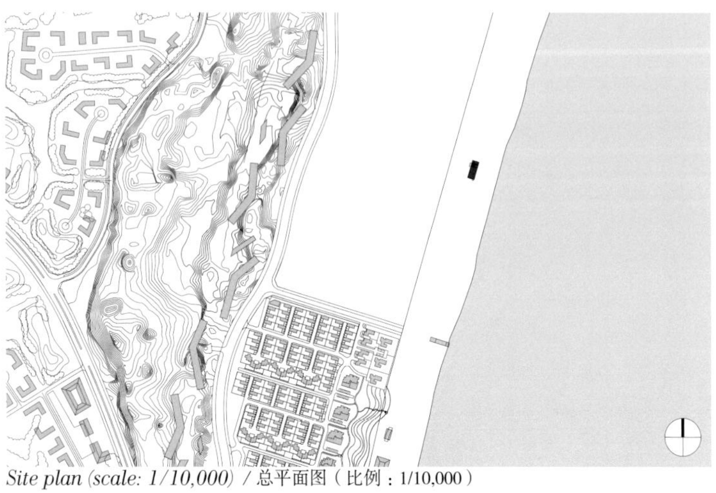

Site plan (scale: 1/10,000) / 总平面图（比例：1/10,000）

Credits and Data

Project title: Seashore Library
Client: Beijing Rocfly Investment (Group) Co., Ltd.
Location: Beidaihe New District, Qinhuangdao, Hebei, China
Design period: February 2014 – July 2014
Construction period: July 2014 – April 2015
Design firm: Vector Architects
Principal architect: Dong Gong
Project architect: Liang Chen
Site architect: Zhang Yifan, Sun Dongping
Design team: Liu Zhiyong, Hsi Chao Chen, Hsi Mei Hsieh
Structural & MEP engineering: Beijing Yanhuang International Architecture & Engineering Co.,Ltd.
Structural consultant: Ji Lixin , Liu Zhongyu
Structure: Concrete structure
Materials: Concrete, laminated bamboo slate, glass block masonry
Building area: 450 m²

建筑与都市
Architecture and Urbanism
Chinese Edition 16:06

064

Feature:
Architects in China

Vector Architects
Seashore Library / Beidaihe New
District, Qinhuangdao, Hebei, China

p. 40: General view from the northwest toward Bohai Bay. p. 41: View through outdoor area on the ground floor. Reading area is on the second floor. pp. 42–43: Elevation facing the sea. A series of glass pivot doors open to connect the reading area and the beach. All photos on pp. 40–47 except as noted by Su Shengliang, courtesy of the architect.

40页：从西北侧望向渤海湾的全景。41页：从一层望向室外空间。42–43页：朝向大海的立面。由玻璃旋转门组成的活动“墙”被完全转开后，可连通阅读空间和海滩。

While Beijing has been experiencing massive growth in the economy and in urban development, many have pointed out the issue of a drop in the living environment. The Seashore Library is one of the facilities in a vacation compound a three-hour drive away from Beijing along Bohai Bay, which aims to create a quality-living area closer to nature.
The design focuses on exploring the co-existing relationship of the space boundary, the movement of human body, the shifting light ambience, ventilation and the ocean view.
The library houses a reading area, a meditation space, an activity room, a drinking bar and a resting area. According to each space, we establish a distinctive relationship between space and the ocean, and define how light and wind enter into each room.

Reading area
By piling up seating platforms raised toward the back, we set the ocean as a stage. A series of glass pivot doors open to connect interior and exterior in nice weather. There is a horizontal window on top that goes across the library, framing the ocean view. To achieve a longer span, all the roof loads are carried by the steel trusses running above the window. Hand-crafted glass bricks are the infill on both sides of the steel trusses, which softens the structure. The translucency bridges natural and artificial light inside and outside throughout different times of the day, smoothly changing the ambience of the building.
The half-arched roof opens toward the sea and shapes the main space while allowing a wide span in both directions. Several 30 cm diameter circular openings are inserted and arrayed on the roof. They are open for ventilation when the weather allows. During spring, summer and fall, from 1:00 to 4:00 pm every day, sunlight streams through those narrow air passageways and projects light spots inside, which meander through space with the shifting of time.

Meditation space
In contrast with the bright and open Reading area, the Meditation space is dim, enclosed and private. There are two 30 cm openings on the east and west sides of the room. The horizontal one captures the sunrise, and the vertical one grasps the sunset. A drastic curved roof pushes the ceiling down low, and creates a low terrace on the roof top. People hear the sound of the ocean here without seeing it.

Activity room
The Activity room is a fairly isolated space. It is separated from the Reading area by an Outdoor platform, due to potential events and noise. Skylights facing east and west collect light throughout the day. Warm light and cold tinted light overlap in the space simultaneously.

If we slice through the building along the long axis, we can see how each space elaborates itself distinctively with the ocean, and how movements and memories of the human body together choreograph a series of experiences.

pp. 44–45: View of reading area. This page: View toward reading area from outdoor platform. Opposite: Interior view of activity room. There are skylights facing east and west. Each space has a different sectional shape, so various spaces are created with different ways to bring natural light into the interior. Photo by Xia Zhi.

44–45页：阅读空间。本页：从室外露台看向阅读空间。对页：活动室内部。朝向东部与西部的天窗。每个空间都有不同的截面形状，所以通过引进自然光的不同方式来塑造各异的空间。

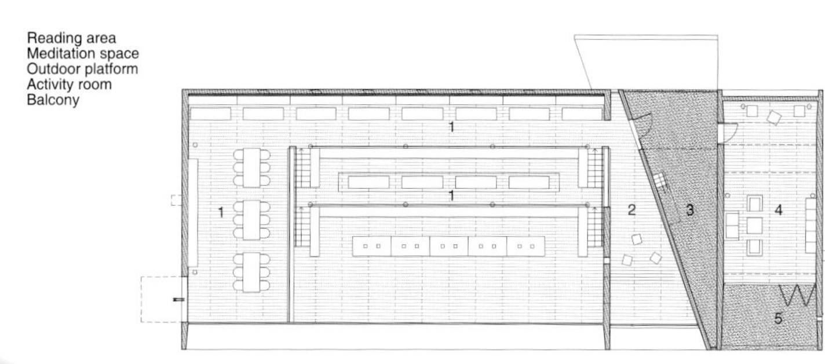

econd floor plan / 二层平面图

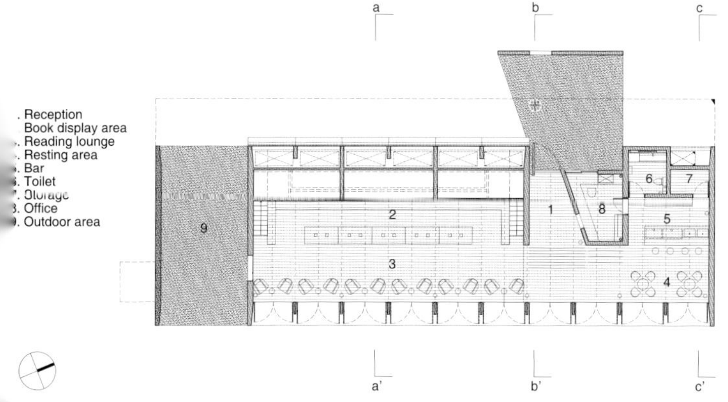

Ground floor plan (scale: 1/400) / 首层平面图（比例：1/400）

随着北京快速的城市发展，生活环境质量的下降已成为日趋严重的问题。这个位于中国渤海湾的社区距离北京约 3 小时车程，旨在提供一处贴近自然的高品质生活环境，本项目是其中的一个文化设施。

设计理念主要在于探索空间界限、身体活动、光氛围变化、空气流通以及海洋景致之间的共存关系。

图书馆由阅读空间、冥想空间、活动室、小水吧和休息空间构成。设计师依据每个空间的功能需求来设定空间与海的具体关系，定义光和风进入空间的方式。

阅读空间

我们将海看作一个戏剧舞台，阅读空间为“看台”，其阶梯平台逐渐向后方层起。天气晴好时，由玻璃旋转门组成的活动“墙”被完全转开，以连通内外空间。“墙”上方，是横贯空间的水平海景视窗。为实现大跨度结构，屋顶完全由视窗上方的钢桁架支撑。桁架内外由手工烧制的玻璃砖砌成墙体，从而柔化了桁架结构的同时，也因这种半透明性对光线的敏感，在一天之中的不同时间，在建筑内外映射出不同的光氛围。

弧形屋顶向海张开，界定了主题空间，也实现了屋顶在东西和南北向上的大跨度结构。屋顶上阵列设置的直径为 30cm 通风井道，在天气允许的情况下可电动开合，使空气流动。在春、夏、秋三季，从下午 1 点到 4 点左右，阳光会穿透这些细窄的风道，在空间中洒下游移的光斑。

冥想空间

相对于阅读空间的明亮和公共性，冥想空间是幽暗、封闭且私密的。空间东西两端各有一条 30cm 宽的细缝。太阳在早晨和黄昏时会透过缝隙在此投射出日晷般的光束。下凹的屋顶降低了空间的尺度，在上方形成一个户外平台空间。在这里，人们看不到海，却可听到海浪的声音。

活动室

活动室是一个相对孤立的空间，考虑到它可能会产生声音干扰，所以将其与阅读空间用一个户外平台区隔开。东西的天窗分别采集一天中不同方向的光，冷暖光同时交替在空间中。

若将这个房子沿南北长向剖开，就可察觉到这一组空间各自诠释着与海的关系，而串联这一系列关系的要素恰是人在空间的游走和记忆。

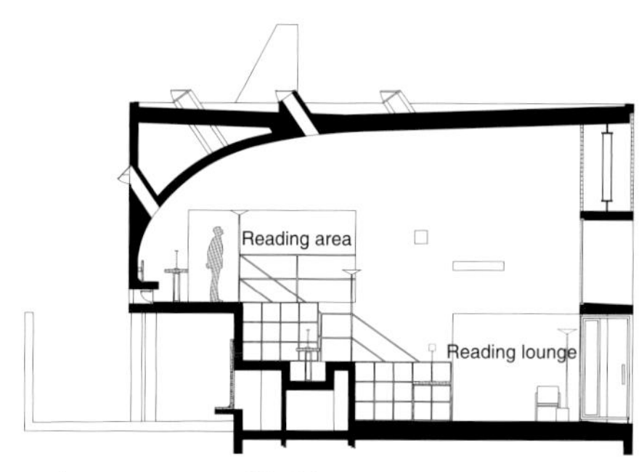

a-a' section / a-a' 剖面图

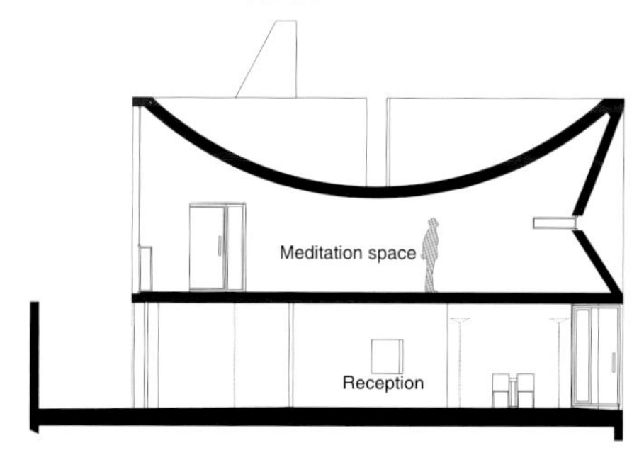

b-b' section / b-b' 剖面图

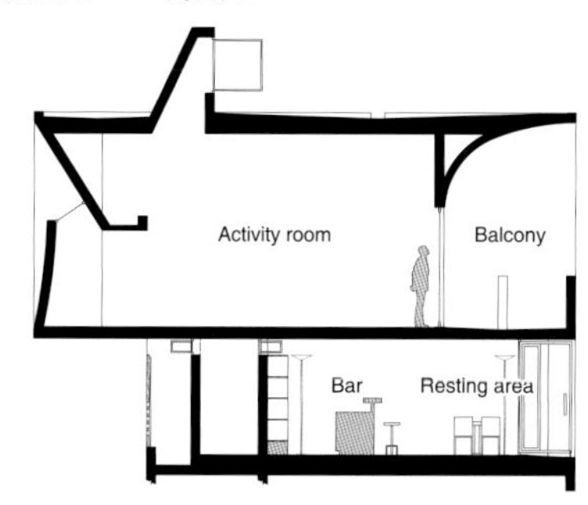

c-c' section (scale: 1/250) / c-c' 剖面图（比例：1/250）

Atelier Z+

Sino-French Centre, Tongji University

Yangpu District, Shanghai, China, 2006

致正建筑工作室

同济大学中法中心

中国，上海市，杨浦区 2006

This page: Exterior view of the college sector, wrapped with COR-TEN steel sheet panels. The ceiling of the corridor has rooflights. Opposite: View of the underground sunken garden. Cafe is on the right. All photos on pp. 48–50 by Zhang Siye, courtesy of the architects.

本页：学院的外部由耐候钢板包裹。走廊的天花板上设有天窗。对页：地下空间的下沉花园。咖啡馆在右侧。

建筑与都市
Architecture and Urbanism
Chinese Edition 16:06
064
Feature:
Architects in China
Atelier Z+
Sino-French Centre, Tongji University
Yangpu District, Shanghai, China

1. Exhibition hall
2. Cafe
3. Preparative
4. Sunken garden
5. Parking
6. Duty room
7. Toilet
8. Storage

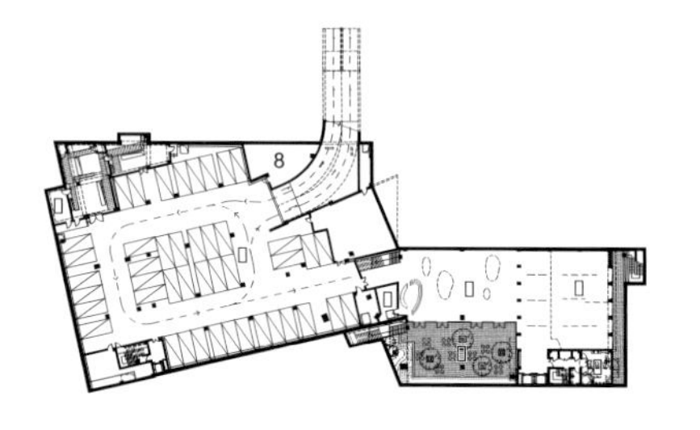

Underground floor plan / 地下层平面图

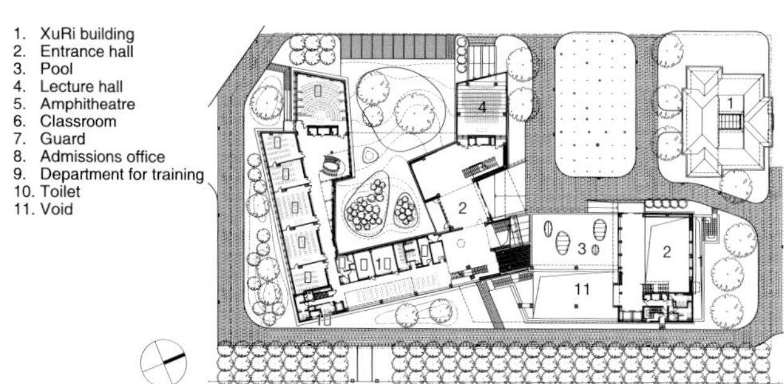

Ground floor plan (scale: 1/1,800) / 首层平面图（比例：1/1,800）

The Sino-French Centre, Tongji University is located at the southeast corner of the campus, with the 12.9 Building, the oldest existing building of the campus, and 12.9 Memorial Park on its west side, a track field on its south side, and Siping Road on its east side. The other precondition is a choice to retain a group of existing metasequoias and nine other scattered trees, such as deodar cedars, plane trees, Japanese pagoda trees and willows. The goal of this project is to create a form system to integrate its program, its site context and its culture context. Our way to achieve it is to use a geometric diagram to control the materialization of its program and circulation, to conform to the site restrictions, and also to indicate its symbolic meaning, the culture exchange between two countries. The diagram of "Hand in Hand" is introduced to organize the whole building with its inherent structure of dualistic juxtaposition.

The program is composed of three parts: college, office, and public gathering space. Two similar but different zigzag volumes, occupied by the college and office sectors respectively, overlap and interlace each other, and then they are linked together by the volume of the public gathering space on the underground and upper levels. The college and office sectors share the main entrance, which is located at the void part of the intersection of these two volumes, while the public gathering space has its own lobby, which faces the roof pool and sunken garden, to connect the underground exhibit hall and the lecture hall on the upper level. Keeping the functions of the college and offices in mind, regular shapes are used for almost all units. Yet applying a zigzag corridor to connect these units creates abundant interest throughout the inside and outside spaces.

Different materials and tectonics are applies to the different components of the complex. The college sector is wrapped in COR-TEN steel sheet panels. Precast cement panels are introduced into the office sector. This treatment indicates the symbolic meaning of this project, the juxtaposition of two different cultures.

Landscape design plays a very important role of this design as wall. The retained existing metasequoias, surrounded by the office sector, public gathering area and XuRi Building at the northwest corner of the site, form an entry plaza of the complex. Connected with 12.9 Memorial Park, this space will become a very important outdoor space to serve the entire campus. The connection between two parts of the building formed a roof pool and a sunken garden, which becomes an intermediary between urban space and campus space.

Eventually, by applying different geometries, materials, colors and tectonics, we create a unique piece of architecture that has a through and profound grasp of the meaning of cultural exchange between China and France.

同济大学中法中心位于校园东南角，西临校园内现存最老的建筑物一·二九礼堂和一·二九纪念馆，南侧为运动场，东侧紧邻四平路。基地南侧的一片水杉林和基地内另有的9棵散落的雪松、梧桐、槐树、柳树等需要保留。

设计师将这一建筑看作建筑形式系统对内部使用功能和外部环境条件及更广阔的文化语境的创造性整合。从项目本身所具有的多层面的“交流性”入手，提出了一个“双手相握”的图解，利用这一图解潜在的“二元并置”结构来组织整个建筑的相关系统，以达成一个“和而不同”的整体。这个图解既是对建筑内部功能和流线系统的抽象，又是源于场地条件对建筑体量的挤压和拉伸，同时也是对中法两国文化的差异并存的关照。

建筑整体分为3个部分，分别用于教学、办公和公共交流。南北两个进深相同、由曲折连廊串联起各使用空间的教学与办公单元互相穿插后，分别从空中和地下结合到公共交流单元。不规则的体量转折和穿插创造了丰富多变的室内外空间，使巨大的体量消解于细腻的环境中，同时大部分使用空间仍保持规则形状。教学、办公单元的共用门厅位于它们上下穿插的虚空部分。公共交流单元另设一个独用门厅，并将地下的展厅、南侧的屋顶水池、下沉庭院和二层的报告厅联系起来。

不同单元采用3种不同的材质组合和构造方法来建构。教学单元采用自然氧化的耐候钢板包裹网格状立面，办公单元用轻质混凝土挂板覆盖立面。这样的两种色彩和材质暗示了中法不同的文化传承的视觉表征。

此外，建筑设计中的景观设计也非常重要。保留的水杉林被办公单元、公共交流单元及旭日楼围合后，成为建筑的入口庭园，并与一·二九纪念馆一起，形成校园中一个重要的公共开放空间。建筑的两个单元穿插处的屋顶水池和下沉庭院既丰富了景观层次，又使建筑本身成为纪念馆空间和四平路城市空间的中介。

最终，通过各种形态、材料、色彩和结构的运用，设计师营造了一个独有的、深刻表征着中法文化交流意义的建筑。

Opposite: Interior view of the stair corridor of the college sector.
对页：学院阶梯走廊的内景。

Credits and Data

Project title: Sino-French Centre, Tongji University
Principal use: College, office, conference centre and exhibition
Location: Main Campus of Tongji University, 1239 Siping Rd., 200092, Shanghai
Design period: March 2004 – October 2006
Construction period: December 2004 – October 2006
Architects: Atelier Z+
Principal architects: Zhou Wei, Zhang Bin
Project team: Zhuang Sheng, Lu Jun, Wang Jiaqi, Xie Jing
General contractor: Zhejiang Hua Sheng Construction (Group) Co., Ltd.
Structure: Reinforced concreteframe, partly steel frame; 1 basement, 5 stories and 1 story penthouse
Primary materials: COR-TEN steel sheet panels, precast cement panels, exposed concrete, steel profile, aluminum, glass, timber
Site area: 9,204 m²
Building area: 3,142 m²
Gross floor area: 13,575 m²
Cost: RMB 60,000,000

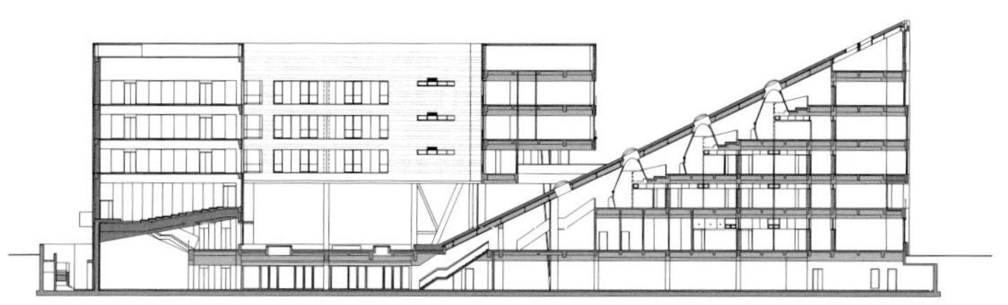

North-south section (scale: 1/1,500) / 南北向剖面图（比例：1/1,500）

Archi-Union Architects

Fab-Union Space on the West Bund

Xuhui District, Shanghai, China 2015

创盟国际

Fab-Union Space

中国，上海市，徐汇区 2015

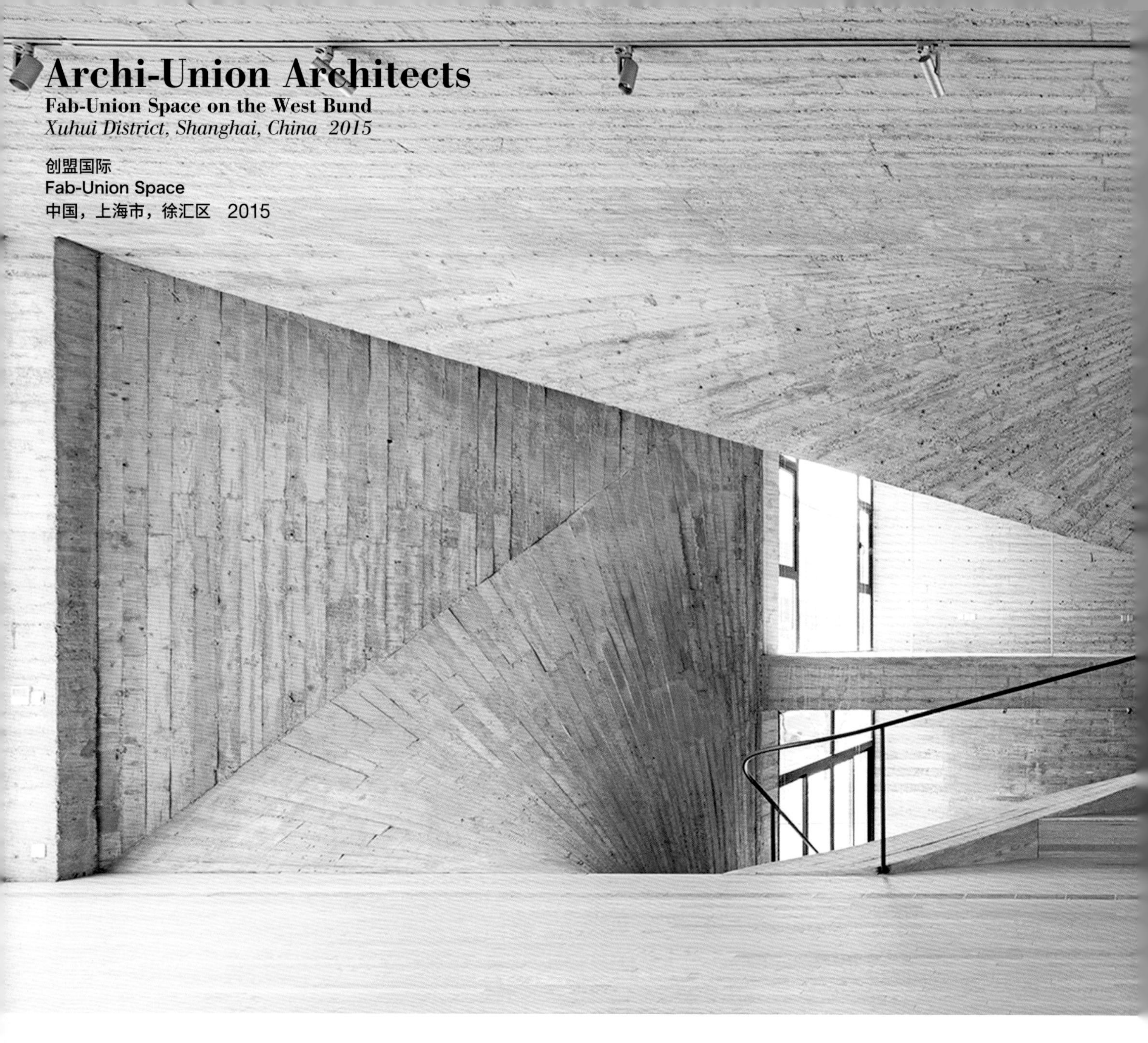

This project, which is intended to be an impressive space in the city, is designed by Philip F. Yuan and Archi-Union Architects. Although it is micro in scale, Fab-Union Space is powerful enough to represent a new attitude to the value shift in architecture. It is located in the West Bund area of Shanghai, which is planned to be a future art and culture centre of the city. Adjacent to Shangh Art Gallery, West Bund Art Centre and several other architecture studios, the location of the site is terrific. It is only 200 m away from Huangpu river front and two blocks away from the historical quarter of Longhua Temple. Moreover, Long Museum West Bund (See pp. 20–27), Shanghai Centre of Photography and Yuz Museum are all within walking distance. It is undoubtedly located within the art community that is making significant experiments in Shanghai. Fab-Union Space will become a future non-profit contemporary art, architecture and culture communication centre. It is intended to be an exhibition and communication space.

The site is very compact, located at a sharp corner. The different circulations from three directions and a connection to the second floor platform of West Bund Art Centre both have to be considered. Therefore, the primary concept is to set up a soft joint for the whole community. The analysis leads to a form finding process throughout the geometry of the site. The inspiration of the material to reach this softness comes from the concrete of the platform, which is originally cold and tough. If we use the mold system, we can actually implement the concrete to any soft surface.

The program is specially set for exhibitions. Five basic spaces – two 4.2-m height spaces and three 2.8-m height spaces – are all regular squares, for multi-functional flexibility. All of the special space experience lies in the in-between circulation space. Interior public space, followed by the exterior form finding process, is enhanced by abstract thinking about creating an experience like climbing the rockery of a Chinese garden. The key aspect of Chinese garden design is to make it big through small scale, which is extremely efficient in changing sceneries with varying viewpoints. The abstract process for this kind of experience is achieved by the hyperbolic paraboloid surface geometry, which is intensified from different perspectives.

The construction process was completed in a very short time. Fabrication marks and traces were recorded on the concrete wall, and a real diversity strengthened a new sense of place in such a compact space. The softness of concrete, which makes the light flow in a leisurely way, deeply touches the hearts of the visitors.

The primary consideration in the conception of this project is to ensure that the exhibition spaces on the side are relatively

complete. But the circulation space of 3 m is built on the basis of the dynamic behavior of people, the air dynamics of the wind and the maximum volume of space continuity.
A dynamic nonlinear spatial shape is built on the basis of structural performance optimization and spatial dynamics. The whole process uses a variety of design methods for cutting stone, perspective geometry, and the algorithm configuration. That the whole building from design to construction required only four months is a miracle of digital design and construction methods.

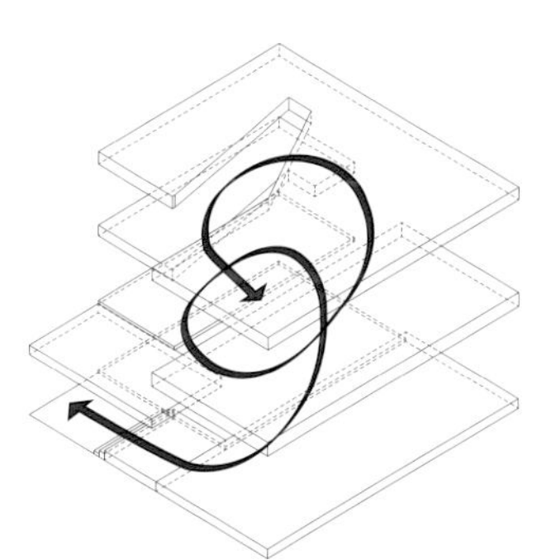

Circulation diagram / 流线图

Credits and Data
Project title: Fab-Union Space on the West Bund
Location: Building D, Longtengdadao, Xuhui District, Shanghai, China
Design / completion: 2015 / 2015
Architectural firm: Archi-Union Architects
Director: Philip F. Yuan
Design team: Alex Han, Kong Xiangping , Wang Xuwei
Structure: On-site cast concrete with steel reinforcement
Structural engineer: Zhang Zhun
Lighting consultant: Hu Guojian , Yang Linhua
Facade Consultant: Shanghai Dimon Curtain Wall Engineering Co., Ltd.
Area: 368 m²

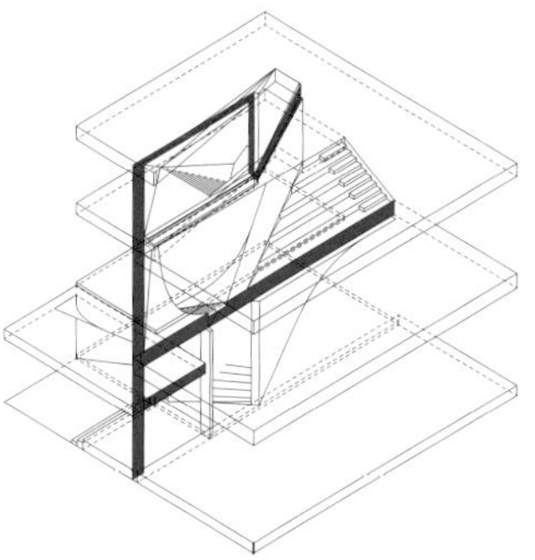

Axonometric drawing / 轴测图

Fab-Union Space 是由袁烽和创盟国际的建筑师为给这座城市增添一处令人印象深刻的建筑空间而设计建造的。它的规模虽小，但强大的空间表现力足以表达一种建筑价值观转变的新态度。位于徐汇滨江西岸文化艺术区的 Fab-Union Space，被规划为城市未来建筑和艺术的中心。它毗邻香格纳画廊的艺术中心和西岸艺术中心，并临近一些建筑工作室，处于城市中极好的地段。距离黄浦江滨仅 200m，越过两个街区即可到达龙华寺，且龙美术馆西岸馆（见 20–27 页）、上海摄影中心和余德耀博物馆都在步行距离之内。它无疑坐落在一个上海重要的实验性艺术社区内。项目旨在打造一个展览与交流的空间，使其成为一个非营利性的当代艺术、建筑和文化交流中心。

基地位于一个急拐角处，位置极其紧凑。设计须兼顾 3 个不同方向的交通流线以及与西岸艺术中心二楼平台的接续。因此，最初的设计理念就是建造一个可使整个街区平缓过渡的建筑。通过对基地形状的分析，建筑形式逐渐生成。使所选材料达到柔性的曲面空间效果的灵感来自混凝土本质的冰冷和坚硬。充分利用模具系统，可将混凝土制作成任何空间曲面。

该项目主要是为展览而建造的。5 个基本空间（其中有 2 个高度为 4.2m，3 个高度为 2.8m）呈方形，并可根据实际情况灵活使用。所有特殊的空间体验都产生于其中的流线空间。建筑外观形式首先被确定，设计师希望在内部可营造一种攀登中国古典园林中假山的氛围，这种抽象设计思维丰富了公共空间的体验感。中国园林的设计要点是“步移景异，以小见大”，而在 Fab-Union Space 中强调的各个角度的双曲面几何形态，就可使人亲切地感受到这种园林体验的抽象过程。

项目施工是在极短的时间内完成的。混凝土墙上留存着无数施工痕迹，它们极真实的多样性强化了这个紧凑空间的“新”感觉。当光线滑进室内，参观者可深切地感受到一种由混凝土产生的温和感。

设计构思首先保证两侧的展厅相对完整，而中间仅有 3m 的交通空间。该构思是建立在人的动态行为、空气动力学拔风以及最大化空间体量连续性的基础之上的。

动态非线性的空间生形建立在结构性能优化以及空间动力学的基础之上。整个过程运用了切石法和投影几何以及算法生形的多种设计方法，从设计到施工历时仅 4 个月，这应是数字化设计以及施工方法带来的奇迹。

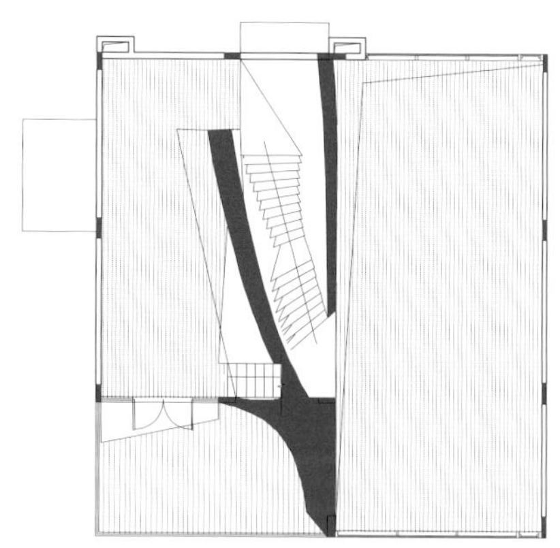

Plan at 3.4 m / 高度为 3.4m 处的平面图

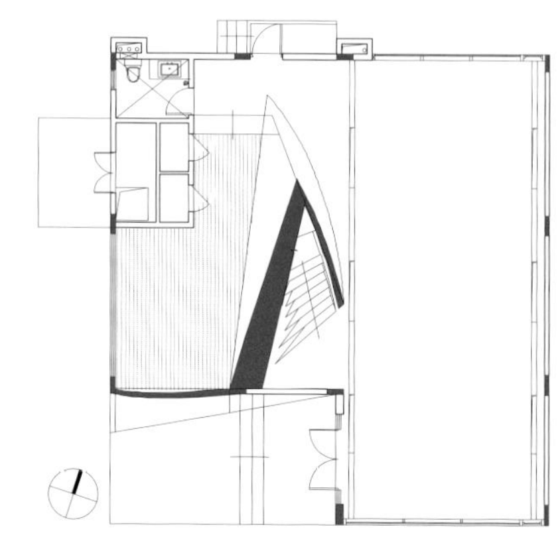

Ground floor plan (scale: 1/300) / 首层平面图（比例：1/300）

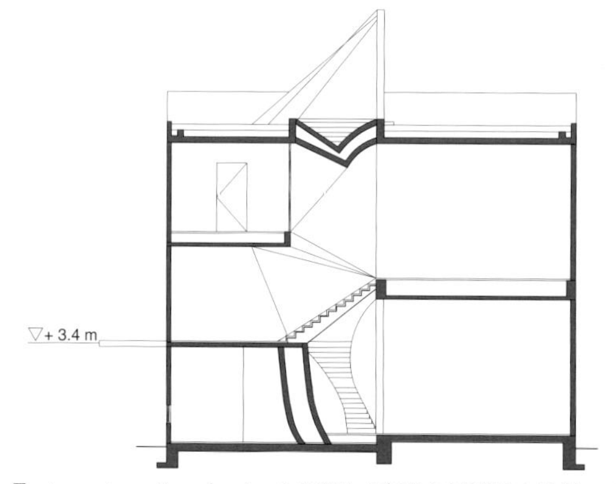

East-west section (scale: 1/300) / 东西向剖面图（比例：1/300）

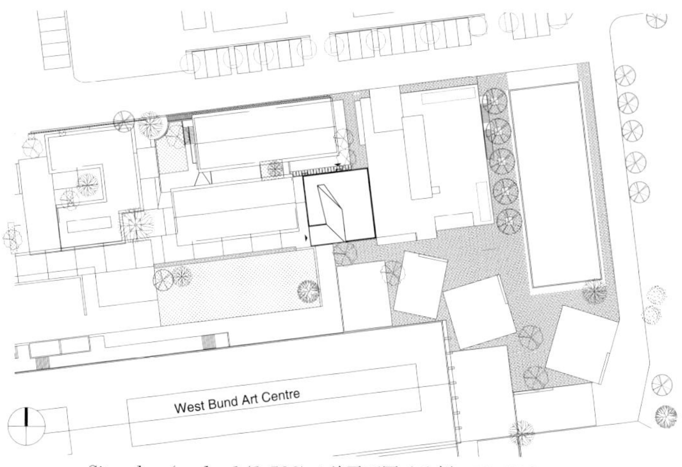

Site plan (scale: 1/1,500) / 总平面图（比例：1/1,500）

pp. 52–53: View of the central stairwell. Opposite: South facade. This page: Interior view of stairwell. The combination of walls creates a space like the caves in Chinese stone gardens. All photos on pp. 52–55 by Chen Hao, courtesy of the architect.

52–53页：中央的楼梯。对页：南立面。本页：楼梯的内景。墙面的组合创造了中国园林中假山的洞穴般的空间。

MAD Architects

Harbin Opera House
Harbin, Heilongjiang, China 2015

MAD建筑事务所
哈尔滨大剧院
中国，黑龙江省，哈尔滨市 2015

pp. 56–57: It is located in Harbin, northern China, close to the Russian border. The wetland is seen in front of the building, and the downtown is seen at the back. Photo by Iwan Baan.

56–57页：项目位于中国北部的哈尔滨，靠近俄罗斯边境。建筑的前方是湿地，后面是市中心。

pp. 58–59: View of the lobby of the grand theatre, finished in Manchurian Ash. People enter from the stairs on both sides which wrap the theatre. All photos on pp. 58–63 except as noted by Hufton +Crow, courtesy of the architect.

58–59页：由西北侧看表面为水曲柳的大剧场大厅。人们由环绕在剧场两侧的楼梯进入。

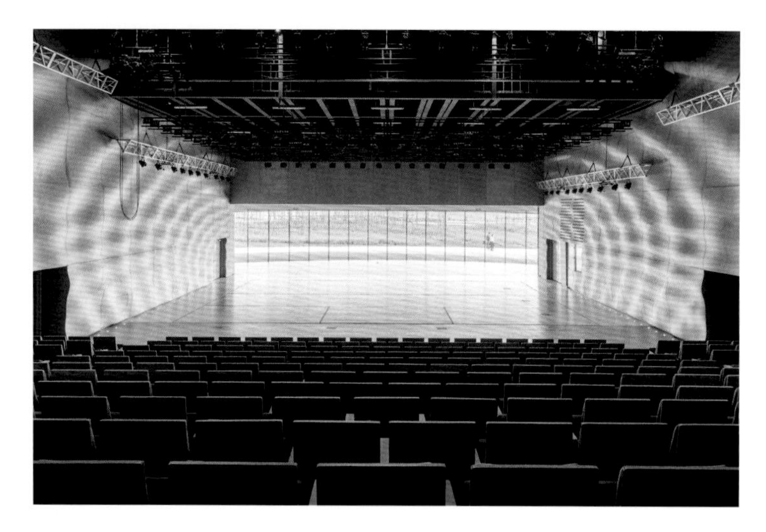

1. Rehearsal room
2. Lobby
3. Grand theatre
4. Small theatre
5. Rehearsal room
6. Entry to parking
7. Stairs to parking
8. Plaza

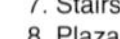

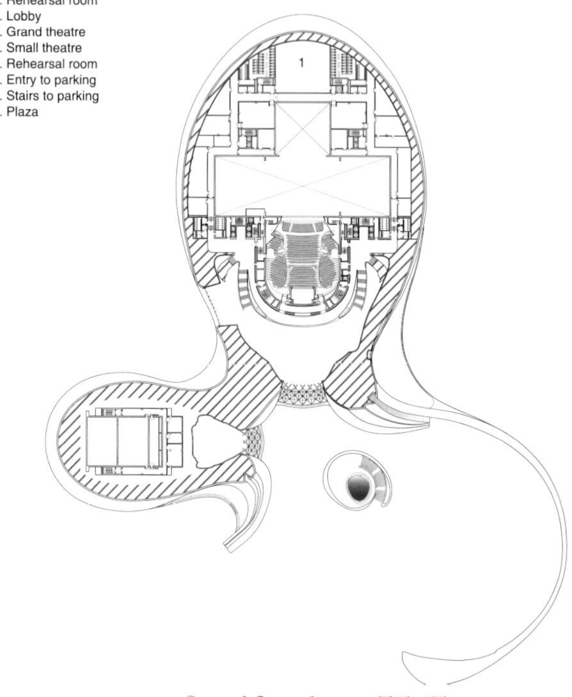

Second floor plan / 二层平面图

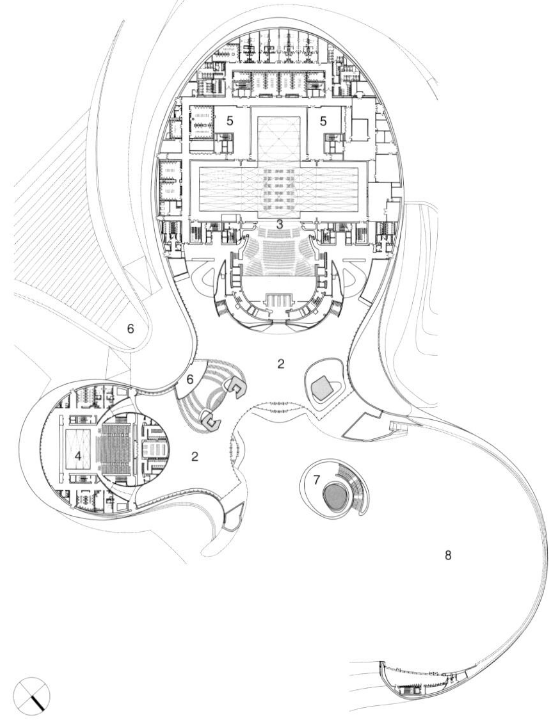

Ground floor plan (scale: 1/2,500) / 首层平面图（比例：1/2,500）

In 2010, MAD Architects won the international open competition for Harbin Cultural Island, a master plan for an opera house, a cultural centre, and the surrounding wetland landscape along Harbin's Songhua River. The sinuous opera house is the focal point of the Cultural Island, occupying a building area of approximately 850,000 square feet of the site's 444 acres total area. It features a grand theatre that can host over 1,600 patrons and a smaller theatre to accommodate an intimate audience of 400. Embedded within Harbin's wetlands, the Harbin Opera House was designed in response to the force and spirit of the northern city's untamed wilderness and frigid climate. "We envision Harbin Opera House as a cultural centre of the future – a tremendous performance venue, as well as a dramatic public space that embodies the integration of human, art and the city identity, while synergistically blending with the surrounding nature," said Ma Yansong, founding principal, MAD Architects.

On the exterior, the architecture references the sinuous landscape of the surrounding area. The resulting curvilinear facade composed of smooth white aluminum panels becomes the poetry of edge and surface, softness and sharpness. The journey begins upon crossing the bridge onto Harbin Cultural Island, where the undulating architectural mass wraps a large public plaza, and during winter months, melts into the snowy winter environment. The architectural procession choreographs a conceptual narrative, one that transforms visitors into performers. Upon entering the grand lobby, visitors will see large transparent glass walls spanning the grand lobby, visually connecting the curvilinear interior with the swooping facade and exterior plaza. Above, a crystalline glass curtain wall soars over the grand lobby space with the support of a lightweight diagrid structure. Comprised of glass pyramids, the surface alternates between smooth and faceted, referencing the billowing snow and ice of the frigid climate. Visitors are greeted with the simple opulence of natural light and material sensation – all before taking their seat.

Presenting a warm and inviting element, the grand theatre is clad in rich wood, emulating a wooden block that has been gently eroded away. Sculpted from Manchurian Ash, the wooden walls gently wrap around the main stage and theatre seating. From the proscenium to the mezzanine balcony the grand theatre's use of simple materials and spatial configuration provides world-class acoustics. The grand theatre is illuminated in part by a subtle skylight that connects the audience to the exterior and the passing of time.

Within the second, smaller theatre, the interior is connected seamlessly to the exterior by the large, panoramic window behind the performance stage. This wall of sound-proof glass provides a naturally scenic backdrop for performances and activates the stage as an extension of the outdoor environment, inspiring production opportunities.

Harbin Opera House emphasizes public interaction and participation with the building. Both ticketholders and the general public can explore the facade's carved paths and ascend the building as if traversing local topography. At the apex, visitors discover an open, exterior performance space that serves as an observation platform for visitors to survey the panoramic views of Harbin's metropolitan skyline and the surrounding wetlands below. Upon descent, visitors return to the expansive public plaza, and are invited to explore the grand lobby space.

As the Harbin Opera House deepens the emotional connection of the public with the environment, the architecture is consequently theatrical in both its performance of narrative spaces and its context within the landscape.

Opposite : Interior view of the smaller theatre. It is surrounded by the glass fiber wall reflecting the natural light and the soundproof glass behind the stage with a view of the wetland. This page: Interior view of the grand theatre with seating.

对页：小剧场的内景。它由可以反射自然光的玻璃纤维墙面和隔音玻璃环绕，从舞台背面可以看向湿地。本页：大剧场座席内景。

2010 年，MAD 建筑事务所赢得了“哈尔滨文化岛”的国际设计竞赛，它是涵盖哈尔滨大剧院、市民文化中心和剧院周围景观湿地的整体规划设计项目。蜿蜒的大剧院是文化岛的中心，其规划用地 1.8km^2，总建筑面积 7.9 万 m^2，由包含 1,600 座的大剧场及 400 座的小剧场组成。

大剧院以环绕周围的湿地自然风光与北国冰封的特征为设计灵感，从湿地中破冰而出。MAD 建筑事务所的创始人马岩松表示：“我们希望作为城市文化中心的哈尔滨大剧院，在拥有巨大表演艺术场地和城市公共空间的同时，也能成为一处人文、艺术、自然相互融合的大地景观。”

建筑的外观设计得益于周边延绵的地域景观。曲线形的外立面采用光滑的白色铝质板包裹，边缘和表面亦柔亦刚，形成了一幅诗意的景域。体验这座建筑，从行走于哈尔滨文化岛之间的桥体开始，缓缓起伏的建筑围合着一个大型的公共广场。冬季，它会融入周围的雪景中并形成另一道风景。

建筑的入口表达了“观众即在场表演者”的叙事性理念。进入壮观的大厅空间，透过通体裹覆大厅的玻璃墙，可看到拥有俯冲立面的曲线状内部空间与外部的广场相接。大厅空间由一个轻量级的斜肋构架支撑，其上部是水晶般的玻璃幕墙。建筑的玻璃金字塔立面，由光滑面与雕饰面组合而成，光滑面以极寒气候地域所特有的冰雪线条为主题曲线形。在此，人们可迎面感受到充溢富足的光和其他自然元素，并由这里走向各自的席座。

建筑整体呈婉约邀请的姿态，大剧场的室内主要以当地常见的木材——水曲柳手工打造，水曲柳墙体环绕舞台和观众席，使氛围柔和温暖，自然的纹理让人感受到空间的生命感。从舞台到包厢，简单纯粹的材料和多变的空间组合为最佳的声学效果提供了条件。大剧院顶部的玻璃天窗最大限度地将室外的自然光纳入室内，自然光洒落在剧场中庭的水曲柳墙面上。在这里，观众既可体验与外界的联系，又可感受时间的流动。

小剧场的后台也设计为透明的隔音玻璃，将室内外连通的同时，使室外的自然环境成为舞台的延伸和背景，为小剧场的舞台创作提供了新的可能性。

大剧院强调市民的互动与参与，建筑顶部的露天剧场和观景平台向市民开放，成为公园的垂直延伸，从这里可以看到松花江江南、江北的城市天际线以及周边的自然景观。即使不进剧场观看演出，市民也可以通过建筑外部环绕的坡道，从周围的公园和广场一直走到屋顶，近距离感受这座穿越绵延起伏地形的建筑所带来的戏剧性体验和意境。顺坡道而下，可返回开阔的广场，探索壮观的剧院大厅。

哈尔滨大剧院是一座可加深人与自然之间情感互动的建筑。也因此，无论是舞台演绎空间，还是在与景观对话方面，它都表达着建筑戏剧性的一面。

Credits and Data

Project title: Harbin Opera House
Typology: Opera house
Location: Harbin, China
Dates: 2010–2015
Architect: MAD Architects
Directors: Ma Yansong, Dang Qun, Yosuke Hayano
Design team : Jordan Kanter, Daniel Gillen, Bas van Wylick, Liu Huiying, Fu Changrui, Zhao Wei, Kin Li ,Zheng Fang, Julian Sattler, Jackob Beer, J. Travis Russett, Sohith Perera, Colby Thomas Suter, Yu Kui, Philippe Brysse, Huang Wei, Flora Lee, Wang Wei, Xie Yibang, Lyo Hengliu, Alexander Cornelius, Alex Gornelius, Mao Beihong, Gianantonio Bongiorno, Jei Kim, Chen Yuanyu, Yu Haochen, Qin Lichao, Pil-Sun Ham, Mingyu Seol, Lin Guomin, Zhang Haixia, Li Guangchong, Wilson Wu, Ma Ning, Davide Signorato, Nick Tran, Xiang Ling, Gustavo Alfred Van Staveren, Yang Jie
Associate engineers: Beijing Institute of Architectural Design
Facade / cladding consultants: Inhabit Group, China Jingye Engineering Co., Ltd.
BIM: Gehry Technologies Co., Ltd.
Building area: 850,000 sq feet
Building height: 184 feet
Grand theatre capacity: 1,600 seats
Small theatre capacity: 400 seats

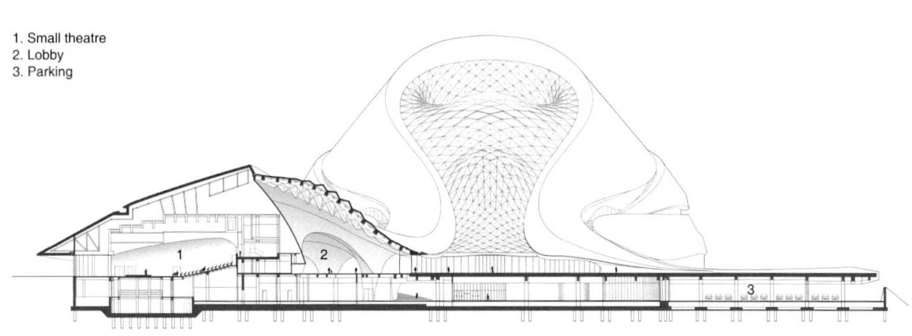

Section through small theatre / 小剧场剖面图

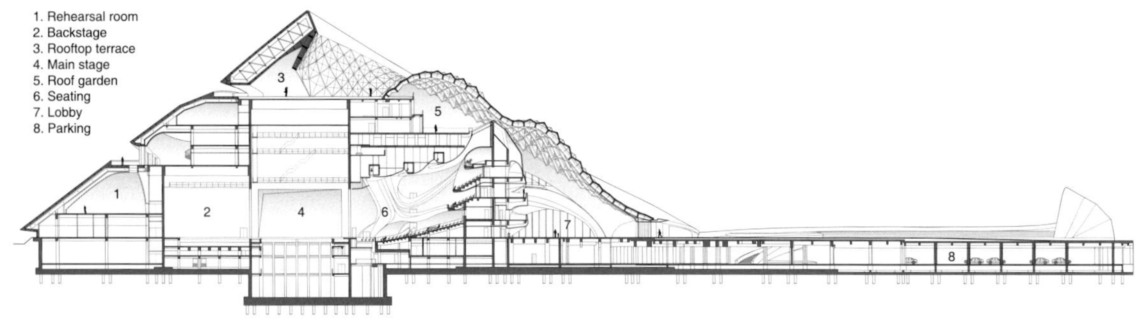

Section through grand theatre (scale: 1/1,500) / 大剧场剖面图（比例：1/1,500）

Opposite, above: View of the rooftop terrace from the north. It is supported by the diagrid structure continuing to the ceiling of the lobby of the grand theatre. There is an observation deck with a view of the city of Harbin and the Songhua River behind the terrace. Opposite, below: Night view looking into the lobby of the small theatre from the plaza. Photo by Iwan Baan.

对页，上：从北侧看楼顶露台。它由与大剧场大厅屋顶相连的斜肋结构支撑。露台后面有一个可以看向哈尔滨市区和松花江的观景台。对页，下：从广场看向夜晚的小剧场大厅。

Studio Pei-Zhu

Minsheng Museum of Modern Art
Chaoyang District, Beijing, China 2015

朱锫建筑设计事务所
北京民生现代美术馆
中国，北京市，朝阳区 2015

The Use of Uselessness – An Attitude of Minsheng Museum of Modern Art
Minsheng Museum of Modern Art (Minsheng MoMA) is a renovation of an old factory from the 1980s. Its openness, diversity, and flexibility make it different from typical galleries, which are always monotonous and secluded. It will be the largest public exhibition space dedicated to Chinese contemporary art.
Rapid urbanization brings us not only a heritage of material civilization, but also a huge amount of waste. The propensity to love the new and loathe the old causes many old buildings to be abandoned. The Panasonic factory at the 798 area is devastated and ruined, but its rough, plain and real industrial building features happen to coincide with the attitude of contemporary art. The concept of Minsheng Museum of Modern Art derives from the above – it respects the simplicity and reality of industrial buildings, behaving harmonically in a completely natural, uncontrived way, use of uselessness, aiming at the future of contemporary art space, challenging the high-sounding traditional art museum.

Diversity of spaces, replacing the single space pattern of "white cube"
Compared to traditional art, a distinguishing feature of contemporary art is the diversity in form of representation. In order to facilitate this, Minsheng Museum of Modern Art not only has the 5 m clear height space of traditional art museums, but also has spaces of different sizes and dimensions: big box, middle box, small box, classic space, courtyard exhibition space black box (multi-function performance, convention, exhibition spaces). They are organized organically by the central space full of tension, combining with open exhibition spaces such as installation park in front of the art museum, exhibition platform on the rooftop, and central courtyard, forming a group of spaces with different dimensions and shapes.

Public and flexible, replacing traditional mode of closure and solidity
Future art museum should be the art space of interaction and communication among the public, art works and artists, rather than the temple of successful artists showing glory. The most meaningful moment of an art work is the interaction and participation of the public, rather than the moment of its completion. Spaces that are flexible, useful, or even useless, can become the motivation of creativity for artists, of specific site and environment, thereby integrating art works, public and art museum as a whole.

Opposite: Interior view of entrance lobby. The volume with a maximum height of about 30 m was inserted into a ruined factory. All photos on pp. 64–67 courtesy of the architect.

对页：入口大厅的内景。高达30m的空间中嵌入了一个废弃的工厂。

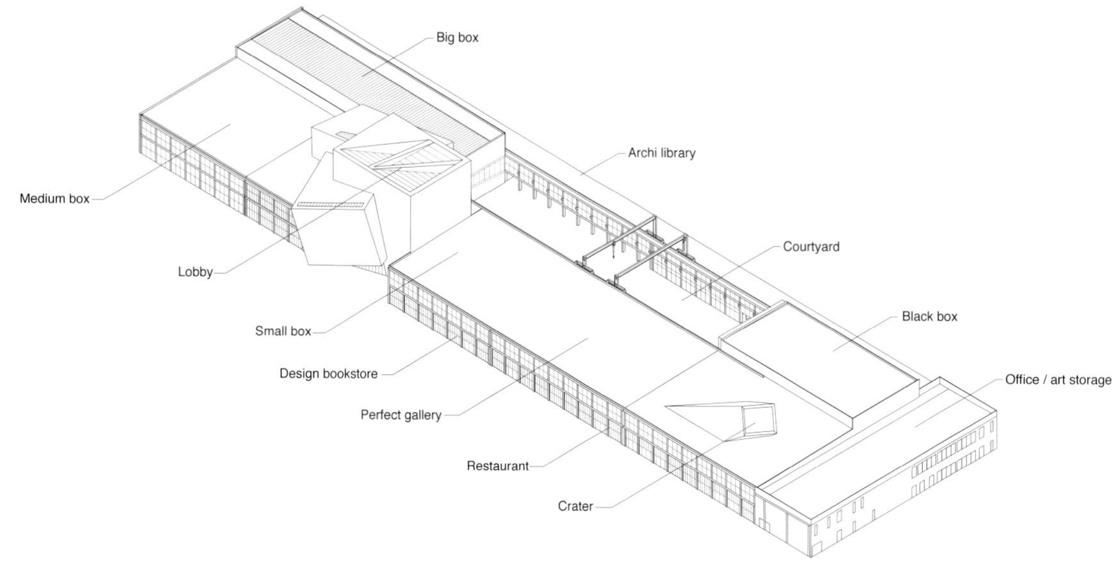

Isometric drawing / 轴测图

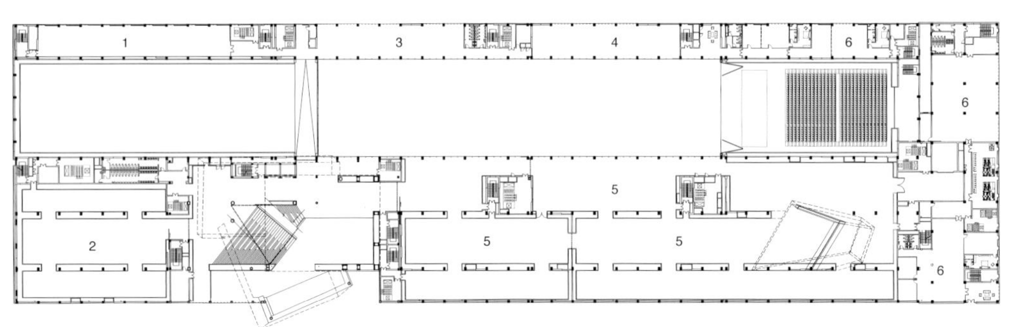

1. Conference
2. Calligraphy gallery
3. Archi library
4. VIP dining
5. Perfect gallery
6. Office

Second floor plan / 二层平面图

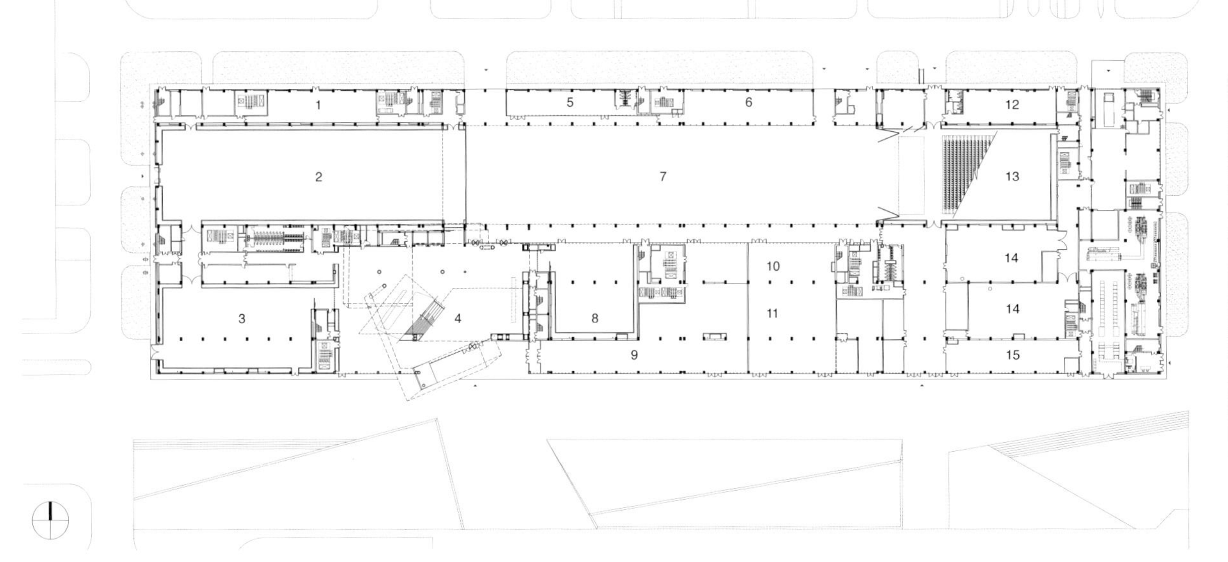

1. Artist studio
2. Big box
3. Medium box
4. Lobby
5. Cafe
6. VIP room
7. Courtyard
8. Small box
9. Design bookstore
10. Art bank
11. Restaurant
12. Dressing
13. Black box
14. Art storage
15. Film shop

ground floor plan (scale: 1/2,000) / 首层平面图（比例：1/2,000）

“无用之用”，民生现代美术馆的态度

该项目由 20 世纪 80 年代的一处工业建筑改造而成。它以开放性、多元性和灵活性，对今天美术馆的封闭性、单一性及固定性提出挑战，将成为中国当代艺术最大的公共平台。

快速的城市化，不仅为我们创造了物质文明的遗产，也为我们身后制造了大量的废弃物。我们喜新厌旧的心理，让众多老建筑遭到遗弃。798 地区的松下显像管厂已不再有过去近 30 年的辉煌，如今遍体鳞伤，满目疮痍，虽破旧、不美观，却透出工业建筑的粗犷、质朴与真实。这些特征恰与当代艺术的态度不谋而合。北京民生现代美术馆正是在这样的基础上诞生的。它尊重工业建筑朴素真实的特质，顺势而为，无用之用，直指当代艺术空间的未来，挑战传统美术馆的冠冕堂皇。

空间多元，替代“白立方”单一空间模式

与传统艺术相比，当代艺术的一个显著特征是其表现形式的多元，为成就这个特征，该美术馆既设有传统美术馆中 5 m 净高的经典空间，也设有大小不一、层高不同的空间：大盒子、中盒子、小盒子、经典空间、院落展览空间、黑盒子（多功能表演、会议、展览空间），来应对不同艺术形式的需求。它们有机地组织在一个充满张力的中心空间周围，结合美术馆前的装置公园、屋顶展览平台、中心院落等开放式展览空间，构成一组尺度不同、形态各异的空间组群。

公共性和灵活性，替代封闭性和固态静止的传统美术馆模式

未来美术馆不再是成功艺术家展现辉煌的圣殿，而是激发公众与艺术作品及艺术家互动交流的艺术场所。空间不再是为呈现作品而作，更是为艺术创作而生。艺术作品最有意义的瞬间不是作品完成之时，而是公众参与并与之互动的时刻。一些灵活可变、有用无用的空间，可激发艺术家和公众为特定环境和场地而创作的激情，让艺术品、公众和美术馆融为一体。

Credits and Data

Project title: Minsheng Museum of Modern Art
Location: Beijing, China
Design: 2011–2012
Construction: 2014–2015
Location: Beijing, China
Architect: Studio Pei-Zhu
Design principal: Zhu Pei
Project team: Edwin Lam, He Fan, Shuhei Nakamura, Damboianu Albert Alexandru, Virginia Melnyk, Guo Nan, Ke Jun, Wang Peng, Li Gao
Museum consultant: Thomas Krens/GCAM
Structure, mechanical and electrical consultant: Arup
Structure, mechanical and electrical design: Shanghai Construction Design and Research Institute Co., Ltd.; BIAD John Martin International Architectural Design Co., Ltd.

Opposite: Front exterior view. The exterior wall of the addition is made of aluminum-magnesium-manganese alloy plates. Photo ©Fang Zhenning. This page, left: Courtyard with preserved elevation of the existing factory. Photo ©Qingzhu Image. This page, right: View of exhibition space leading to the entrance lobby by a grand staircase.

对页：正面外观。增建部分的外墙由铝、镁、锰合金板构成。本页，左：保留了原有工厂外立面的中庭。本页，右：通过一个宏大的楼梯连接入口门厅的展览空间。

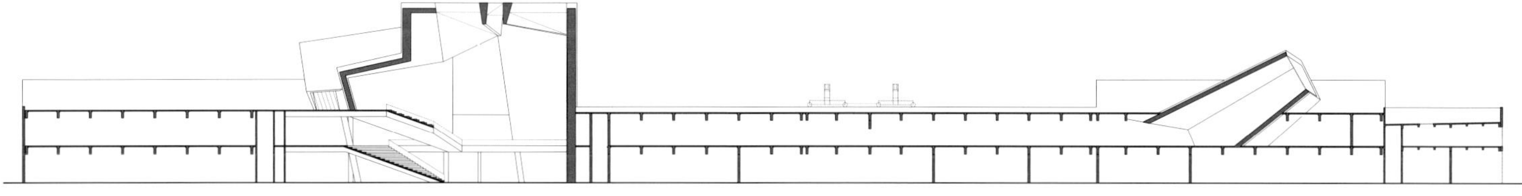

Longitudinal section (scale: 1/1,500) / 纵向剖面图（比例：1/1,500）

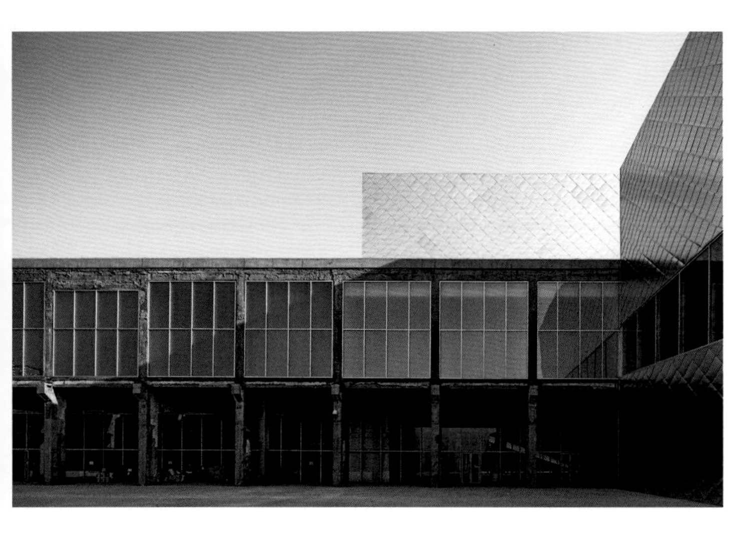

Discussion:

Regional Development of Architecture

Zhu Pei, Zhang Ke, Tao Lei, Wang Shuo
Moderators: Ma Weidong, Li Xiangning

对谈：
建筑的地域性发展
朱锫 张轲 陶磊 王硕
主持者：马卫东 李翔宁

—In this discussion in Beijing, four architects were invited: Zhu Pei, Zhang Ke, Tao Lei, and Wang Shuo. They discussed modern Chinese architectural development from the area of focus of each architect. Having their base in Beijing, and born in the 1960s, 1970s, or 1980s during a special age when the Chinese architecture industry developed, they have experienced the development of architecture in Beijing in various forms. In the interview, they talked about a "Pilgrimage of the Mind", "Beijing Viewed from an Architects' Perspective", and the "Current Situation of Architects and Construction Technology". At the end of the interview, "Regional Characteristics", an always important theme for domestic and international architects, was discussed.

Pilgrimage of the Mind

Zhu Pei (ZP): I was born in the 1960s, and came back to China in 2000 after studying and working abroad. At that time, architects in the older and younger generations were actively working, and I felt quite uncomfortable. The older generation architects included those who began their independent careers in the 1990s, Wang Shu, Yung Ho Chang, Cui Kai, and Liu Jiakun. They had already begun to create rather experimental architectural works. For a while after returning to China, I was hoping to change the situation by utilizing the design philosophy and construction skills I had learned abroad. However, when looking back, I think I was dependent on foreign customs and work experience, I had a strong yearning for the simplicity and experimental features of modernist architecture, and I was biased toward my own personal preference, minimalism. Apparently I intentionally avoided traditional building materials and construction methods. After completion of my first work, "office building in Shenzhen Planning Bureau", in 2004, I began to think about how to entrench architectural works into a specific culture or region.

Zhang Ke (ZK): My first project after returning from New York in 2001 was a cultural project, the "Ming City Wall Relics Park" in Beijing. Due to this project, I began to become directly involved with the history of the city of Beijing, and our practice also became interested in and focused on the history of Beijing. This had been a continuous theme since my undergraduate studies at Tsinghua University. I have a kind of plot about the culture of Beijing. When I returned home, I was quite rebellious and felt that architects at that time were lacking a sense of mission. I was enthusiastic to show myself succeeding in a fine project. I was highly self-conscious and refused to imitate architectural styles in Europe, the U.S., or Japan. I was thinking of creating China's original style, not only exterior but also interior designs. A direct embodiment of China evokes laughter, but characteristics of China that I currently pursue are underlying things, realistic things obtained through experiments.

Tao Lei (TL): I never studied abroad and began my career in 2002. In Chinese architecture at that time, genuine modern architectural works were really rare, though new conceptual architecture practices were already beginning by some architects, including Zhu Pei. At the same time, a new and highly perfected concept was developed in architecture by adding a standardized manipulation process. Such a practice can be described as the arousal of Chinese culture, and I began to have some sort of confidence in the beginning of a new value in Chinese society. I never actively thought of characteristics of China, but I believe that customers' needs and judgement naturally lead to characteristics of China. Therefore, if we become faithful to realistic problems and our feelings, and add new thoughts and plans, we will be able to obtain our own uniqueness. I think such uniqueness only exists in given conditions in architectural projects.

Wang Shuo (WS): I finally went outside China in 2004, and I personally regard this as an "exploration" experience. Beijing is my hometown, and I continued exploration of "what is the quality of life in this city". Since then, I have conducted urban research, "reconstruction of urban stocks", on migrant worker communities around Dahongmen, Beijing.
I am fond of spaces that "are confused, foreign, and cannot be quantified, but have a unique quality of life", such as Hutong in Beijing, and hope to maintain this type of "quality". Through a study of current records, I would like to understand the "source" of things upwelling in the city, explore the systems in the background, and finally lead to a new policy. My job is related to the understanding of such spaces. This triggered me to focus on hollowing out the urban centre, and a broad range of research called [META:HUTONGS] (pp. 72–74) was launched to study Hutong, located inside of the 2nd Ring Road at the very centre of Beijing, in terms of sociology, psychology, new media, policy, and development planning.
As an architect, we must see places where new things are born. We need to find the original point of social problems, rather than having a simple criticism. Excellent designs can solve or mitigate contradictions (such as a situation where a billionaire and homeless person drink coffee together at the same table), resulting in the creation of new possibilities. I think they finally become perfect when a more effective way to solve urban

——本次北京对谈我们邀请到朱锫、张轲、陶磊、王硕四位北京建筑师，分别从各自专注的领域谈一谈当代中国建筑的发展。四位建筑师出生于（20世纪）中国建筑发展的三个特别时期：60年代、70年代、80年代，他们均立足于北京，对故土北京的建筑发展有各自深切的体会。对谈从“心路历程”开始，逐渐过渡到“建筑师眼中的北京”、“建筑师的生态与建造技术”，最终落点于国内外建筑师们热议的“地域性”问题。

心路历程

朱锫：出生于60年代并有国外留学和工作经历的我，在2000年回国后处于“前有猛虎后有追兵”的尴尬位置。所谓猛虎是90年代就开始独立实践的建筑师——王澍、张永和、崔凯、刘家琨等，那时他们已开始独立地进行一些具有一定规模的实验建筑的建造。刚回国时，我主要靠的是国外留学和工作经历的惯性，希望能利用这些惯性在设计观念和建造技术上寻求突破和颠覆。如今反观，这个过程充满了惯性和现代主义建筑的单纯性与试验性，也带有个人极简主义的偏好，建造上有意回避采用传统的建筑材料和建造方式。当2004年完成了自己的第一个作品“深圳规划局办公楼”之后，才开始思考建筑如何根植于特定的文化和地域，跨越惯性。

张轲：2001年，我从纽约回国后的第一个项目是北京的明城墙遗址公园。它是个文化项目，但直接与北京城市的历史发生关系，事务所从那之后，开始对这方面的项目感兴趣，从兴趣又逐步发展为关注。这其实延续了我从清华毕业时的研究，我对北京文化有一种情结。回国后我对当时的情况十分反叛，觉得当时的建筑师使命感不强，我们要做些好的项目给大家看看。自我意识比较强，拒绝模仿欧美、日本风格，要做中国自己的东西，而这些东西不是一眼能看出来的，因为具象的中国大屋顶下的东西是可笑的。现在我追寻的中国性是潜在的，并且是通过实验获得的，是现实的。

陶磊：我没有出过国，从2002年开始工作。当时中国建筑很少有真正意义上的现代设计，但朱锫等建筑师就已经开始进行了一些新概念建筑的实践，并有了更加规范化的操作流程，建筑有了新的概念，完成度也很高。他们的实践似乎是一种中国文化的苏醒，是中国社会新的价值观的开始，给了我一种莫名的信心。关于中国性的问题，我并不是主动去思考的，我觉得客户的需求和他们固有的好恶判断中自然地带有中国性。所以我们应忠实于实际问题和自己的情感，并可以有新思维和新策略面对，这样一定会有自身的独特性，这种独特性只存在于项目的条件中。

王硕：我2004年才出国，所以个人的经历，其实更是“寻找”。北京是我的家乡，我一直在探寻这个城市的生活质量到底是什么。后来我就做了北京大红门的外来城市人口产业聚集区的城市研究，这个现在叫“城市存量改造”。

我对混乱的、异质的、无法量化却又有独特生活质量的空间，比如北京的胡同，有一种原生的喜好，想延续这种质量。我设想通过对当下现实的记录研究，了解城市中涌现的“物种”，找到其背后的运行机制，并在此之上导出新策略。我的工作一直与这个理解有关。继而我们开始发现了城市中心的空心化。于是，又发起了对北京最核心的二环以内的胡同的跨学科研究——[超胡同]，它涉及社会人类学、心理学、新媒体、政策及开发策略等多方面。

作为建筑师，要看新东西产生的地方，我们不仅要单纯地批判，更要找到社会问题的发生点。好的设计能成功化解或缓和对立的矛盾，比如一个街头咖啡馆，能让一个百万富翁挨着一个流浪汉一起喝咖啡，于是产生了新的可能。我的想法是，找到更有效的方法去化解城市中的矛盾冲突（urban conflict），将研究转化为设计，实现后的效果都具备了才算完整。

建筑师眼中的北京

朱锫：我们从小生活在北京，身上都有北京无形的烙印。北京是皇城——皇上在上，百姓自傲，于是文化带有很强的反叛性和幽默性，有一种自嘲的意味。此外，北京一直崇尚文化，有驾驭文化的想法，玩儿什么都讲究玩儿得好，像蛐蛐、鸽子、毽子等，这种玩的心态俗气却很真实。60年代的我们始终觉得建筑不只是一种职业，不应该是被动的，应该是在身心自由的情况下做出来的，应该有一种类似艺术家的玩儿的心态。

我刚从国外回来的时候，第一追求质量，有种典型的现代主义的态度；第二追求艺术，以艺术家的方式从事创作——注重建筑自身的价值与韵味，这个价值就是给人提供一种未曾经历过的体验，给人带来惊喜。这与北京古老的“玩”的艺术和中国古典文化中庄子“自然无为”的哲学观可以联系在一起，“人生充满自由和艺术，应很欣

conflicts is found and translated into a design, and then they will demonstrate its effect.

Beijing Viewed from an Architects' Perspective

ZP: We were born and grew up in Beijing, and "Beijing" is deeply ingrained in us. Since emperors historically lived in Beijing, citizens have pride and their culture heavily reflects self-deprecatory rebellion and humor. There is a culture to perfectly master play (such as enjoying cricket sounds, pigeons flapping, and Jianzi or shuttlecocks), and people tend to exploit this culture. We, born in the 1960s, do not consider being an architect as a job. Rather, with the idea that we can conduct architectural works actively when our mind and body are free, architects are similar to artists who enjoy their work.
The thing I firstly sought after returning home was quality, and this was a typical modernist attitude. Then I pursued art. What I valued was creation as an artist, in other words, value and elegance of a building itself that provides unprecedented experiences and surprises. This can connect the art of play in Beijing with a philosophy of "non-action and naturalness" in the Chinese classic "Zhuangzi". It is possible to say that if life is filled with freedom and the arts, people can enjoy getting things going. Such a culture in Beijing with a rebellious nature and humor is clearly different from the culture in Shanghai. In simple terms, Beijing culture is "rough culture" and Shanghai culture is "sophisticated culture".

WS: I had a similar experience as Zhu Pei's. Right after returning home, I was putting rather too much pressure on myself, and was keen to clarify everything. Later, I realized there are things that can be proved not only by drawings and logic, but also by "aura". Therefore, I am ambitious to transmit this experience to everyone through my research: if the surrounding environment is different, the aura is also different.

TL: Beijing is intricate. "Conservative" and "modern" coexist and sometimes conflict. But I find this conflict most interesting.

ZK: Sometimes things are "unclear" in Beijing, but it might actually be close to the essence.

ZP: Architects in Shanghai are more delicate and value construction works, but those in Beijing might focus on the concept of play, and consider "why it exists here".

ZK: Today, differences between Beijing and Shanghai are significantly greater. For example, when the client is government officials or a developer in Shanghai, everyone will have meetings to discuss technical issues such as air conditioning, the façade, and materials. However, people from the northern part of China, including Beijing, never discuss technical issues, but only philosophy.

Current Situation of Architects and Construction Technology

ZP: Balancing and paying attention to the whole are important. The decision whether to focus on supervising at a work site or on drawing more plans is made by clarifying the overall structure of the client organization. I only visited the work site twice for the OCT-LOFT construction project, and a few times for the Minsheng Museum of Modern Art (See pp. 64–67), but both construction works went smoothly. The biggest problem was how to obtain a desired (i.e., following my design) working drawing from the Design Institute, of which the client outsources. Therefore, taking into consideration the overall structure of the client organization, I decided how to make my design.

WS: I feel that a highly comprehensive capacity is required for architects. From casual communication with clients to details such as the color of screws, architects must manage by themselves.

ZK: It can be said that Chinese architects are playing the roles of carpenters in the past.

WS: Another important point in architectural practice in China is a necessity for considering tolerance in a design. If a little margin for adjustment is left at the design phase, errors at a construction site can be tolerated and mistakes can be adjusted for, resulting in a higher satisfaction at the end.

TL: I think so, too. If no tolerance is considered at the design phase, the project itself might become unfeasible due to various errors.

Regional Characteristics

ZP: In our subconscious mind, we think about how to manifest regional characteristics in buildings: in other words, "how to solve our own problems" and "how to show the world the modern Chinese architecture". Simple integration of regional characteristics and Chinese traditional elements is practiced and is a hot topic everywhere.
Every land and its residents have different living habits and culture, and a certain natural environment, geography, and climate nurture unique lifestyles, cultures, and architecture. We should respect nature's laws as well as the geography and climate around the construction site. Without taking these elements into account, things will be meaningless even if domestic materials are used.

WS: I think architects should consider local geographical nature, but in a subtle way. In terms of the media, regional works were rarely seen on the internet in the past. Today, internet media such as ArchDaily is well developed today, and designs become instantly global. Meanwhile, we emphasize regional characteristics as if using a Chinese cup for drinking coffee.

ZP: For Chinese people who like drinking tea, the tea leaves are more important than the cups. It is closer to natural conditions. I feel sympathy for the "cup theory", but the sympathy fades if the liquid in the cup is not tea.

ZK: The essence is the same as tea. Traditionally in China, pursuit of perfection underlies as an essence, considering integration of architecture and nature.

ZP: That is right. Confronting nature with good faith, and building with minimum energy have significance. This is just like the Chinese traditional philosophy of "non-action and naturalness", that is, a philosophy that considers that the heavens and human beings are united, and follows the course of nature. Going back in time is impossible, but history tells us the essence of architecture: nature creates architecture. In other words, what creates architecture is not human beings.

Translated from Chinese by Haruki Makio

赏、愉悦地做一件事”。这种反叛性和幽默性的“京味”文化与上海文化显然不太一样，简单地说就是北京“糙”，上海“精”。

王硕：我也有与朱培相似的体会，刚回国时比较紧张，格物致知，希望把所有事情说清楚。后来我发现有的事情不光靠图纸、靠理论证明，还有一种气韵（Aura），所以我们通过研究向大家传递一种经验。每个设计的周边环境不同，气韵就不同。

陶磊：北京较为复杂，“保守”与“前卫”同时存在，甚至会发生冲突，这种冲突也是我认为最有意思的。

张轲：北京虽有时不靠谱，但可能更接近本质。

朱锫：上海建筑师的思维相对比较精细，认为建造是一件很重要的事，北京建筑师则会以玩的观念，考虑“我是为什么存在的”。

张轲：我觉得北京与上海的差异至今仍很大，比如上海的甲方，包括政府、开发商，都会和你坐下来谈空调、幕墙、材料等技术问题，而北京乃至整个北方都很少有人跟你谈技术，都在谈思想哲学。

建筑师的生态与建造技术

朱锫：我觉得这是一个短板效应。你必须弄清业主的整个体系是什么样的，这将决定我应该多去现场还是多画图。“华侨城创意中心”项目，我只去了两次现场，“北京民生现代美术馆”的工地，也只去了几次，但施工质量都很好。最大的困境就是如何让业主和合作设计院按照你的设计完成施工图。所以我在考察了体系之后，决定我的施工图怎么画。

王硕：我觉得对建筑师来说，综合能力要求很高。从和甲方喝茶到工地上的螺钉是否上色，都需要自己管。

张轲：这正好反映了中国建筑师的实践是比较接近古老工匠的一种实践。

王硕：中国实践的另一个重要问题就是设计中还须考虑到容忍度，我们在做设计时，如果考虑好这点，能接受施工有误差，有错误能调整，那满意度就比较高。

陶磊：我也同意这个观点，如果设计时没有考虑到这个问题，那么因为各种误差的存在，会直接导致项目无法实现。

地域性

朱锫：我们潜意识里都在思考地域性如何在建筑中呈现。如何解决自己的问题？如何让中国的当代建筑给世界带来启发？“地域性”很容易就和中国的传统元素联系在一起，这是大家都在谈和做的。“一方水土养一方人”，特定的自然环境或地理气候会塑造一种特定的生活方式、文化和建筑。我们须尊重自然法则和当地的地理气候，若你不考虑自然法则，使用再多的本土材料也没有意义。

This page: Image of multiple layers of Hutong socio-cultural space, from Wang Shuo's research project [META:HUTONGS]. Image courtesy of the architect.

本页：胡同里的各种社会文化空间。资料来自王硕的[超胡同]研究项目。

王硕：我觉得建筑师考虑本土性的地域，应是潜移默化的。从媒介的角度谈，现在由于ArchDaily等网媒的发达，设计变得瞬间全球化，我们反而会强调“地域性”，比如同样在喝咖啡，但我们使用的是中国的杯子。

朱锫：对中国人来说，如果你喜欢喝茶，那茶比杯子重要。它符合你原始的状态。我对“杯子”的说法有很强的认同感，但如果杯子里不是茶，事实就变了。

张轲：秉性是茶。中国传统的秉性上有对“完美”的追求，会考虑建筑和自然的融合。

朱锫：对，诚恳地对待自然，用最少的能源材料来建造，这更有意义，也契合了中国传统文化中最重要的“自然无为”的思想——天人合一，顺势而为。我们无法回到自然的远古，但远古告诉了我们建筑的本质，即自然创造建筑，而非人类创造建筑。

META-Project

[META:HUTONGS]

Beijing, China

META-工作室

[超胡同]

中国，北京市

HUTONG ADAPTATION - Perpetual Dynamism

HUTONG COMMERCIAL EVOLUTION - THE TIMELINE

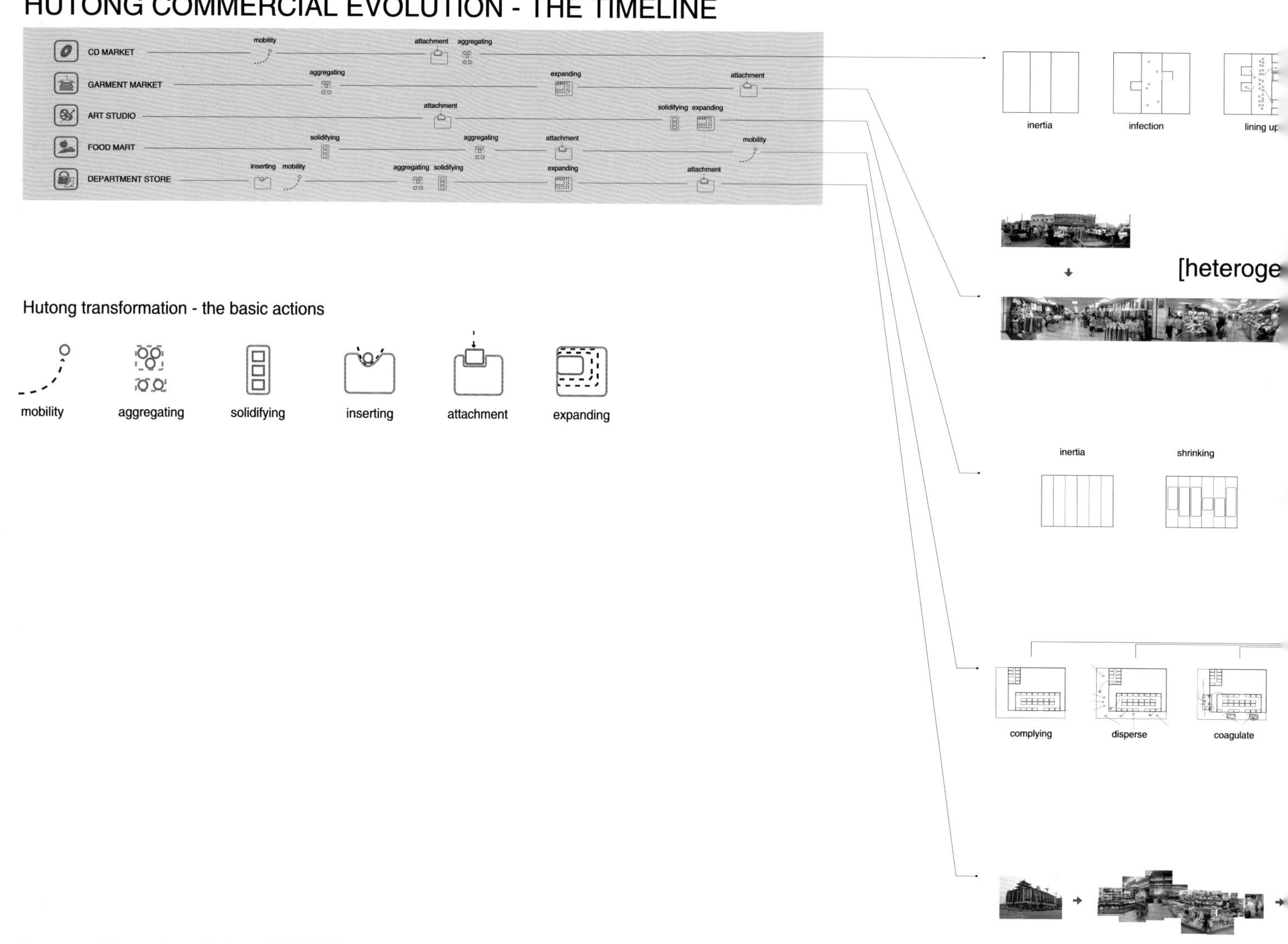

Diagram of Hutong's evolution / 胡同进化图

旭太奇服饰
华生
服装百货店
陈迎时装店
Axin
阿欣时装店
中国书店
CATHAY BOOKSHOP
华都靓装

隆福寺
商店
靓丽装苑
服饰中心
LOLO
隆福商业

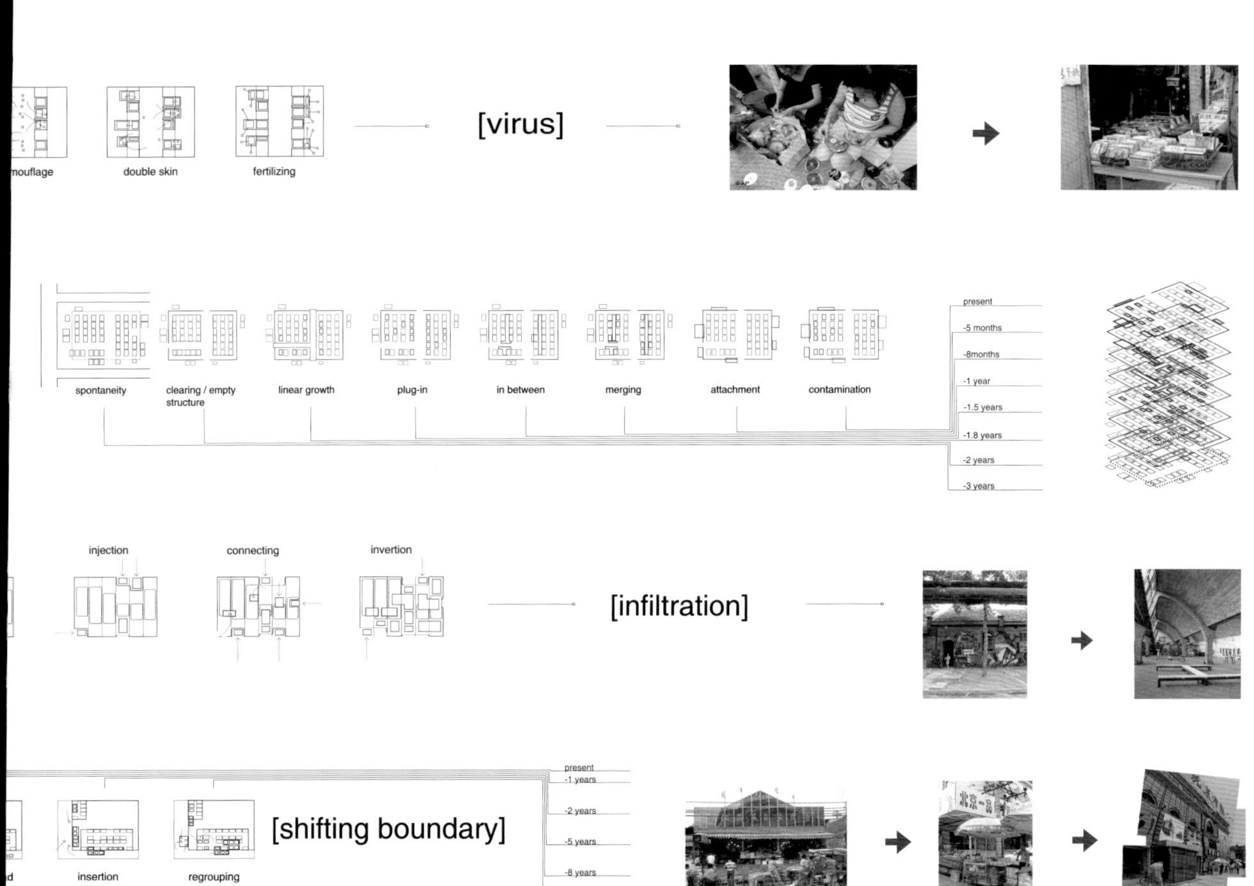
mouflage
double skin
fertilizing
[virus]
spontaneity
clearing / empty structure
linear growth
plug-in
in between
merging
attachment
contamination
present
-5 months
-8months
-1 year
-1.5 years
-1.8 years
-2 years
-3 years
injection
connecting
invertion
[infiltration]
insertion
regrouping
[shifting boundary]
present
-1 years
-2 years
-5 years
-8 years
-10 years

[invert explosion]

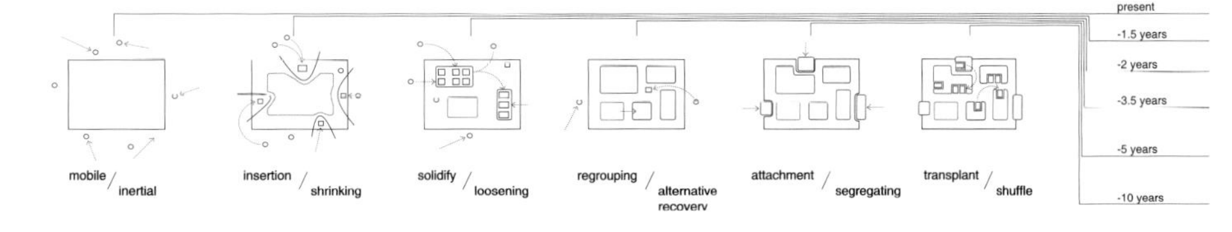
present
-1.5 years
-2 years
-3.5 years
-5 years
-10 years
mobile / inertial
insertion / shrinking
solidify / loosening
regrouping / alternative recovery
attachment / segregating
transplant / shuffle

Hutongs

The narrow alleyway connecting small courtyard residences, or Hutong, has been heralded as the definitive exemplar of Beijing's urbanism since the city became the capital during the Yuan dynasty. To this day, legions of critics and writers have looked to Hutongs as a kind of architectural endangered species, evidence of how traditional urban fabrics have been crushed in the maw of modernization. This explanation is too convenient, too simple, however, and does not capture the reality of the situation in Beijing, a city that has been enduring massive political and social upheavals in ways others could not, one that continuously defies categorization or explanation.

[META:HUTONGS]

Born in Beijing, and now based in the centre of the old city's core, Wang Shuo has been continuously concerned about the Hutongs' evolution and regeneration.
Since 2012, together with his friend Andrew Bryant, he initiated the [META:HUTONGS] project, which became an award winning, internationally recognized research collaborative.
It has been sponsored by Graham Foundation and the Urban China Initiative (UCI) over a period of three years.
[META:HUTONGS] is a collaborative platform that examines, reveals and envisions urban development as it unfolds in the Hutongs. It consists of a series of workshops, charrettes, symposia, and roundtable discussions that consider Hutong as a unique socio-cultural space, a locus of distinct urban data, as well as an exemplar of urban regeneration. Architects, planners, media artists, social anthropologists, and historians will all participate in this investigation with the goal of providing alternative models for decoding and instigating one of the world's most dynamic and vital cities-Beijing.
At this moment, Hutongs have become laboratories that exemplify the current reality of Beijing's urban situation and are a key to potential future urban regenerative strategies.
Public discussions in Beijing have become increasingly obsessed with the utopian idea of preservation, and yet pressure from developers is prompting massive demolition and rebuilding from a tabula rasa. In this relentless, dystopic reality, Beijing's Hutongs are disappearing at a rapid pace with no alternative models for effective regeneration. Between the UTOPIA and DYSTOPIA, [META:HUTONGS] looks at the current reality. The goals of this cross disciplinary investigation are to generate dialog for understanding and evaluating the unique qualities of this urban emergence, and to provide alternative models for instigating projective possibilities.

胡同

那些连接着四合院的狭窄巷道——胡同，是从北京成为元大都时起就形成的鲜明的城市特征。迄今为止，大量的评论家和作家都将胡同视为一种即将灭绝的建筑“物种”，和“传统的城市肌理是如何夹裹在现代化的造城运动中被蚕食”的见证。然而，我们认为这样的解释太过方便，也过于简单，它无法捕捉到北京城真实的现状——一个曾经历过许多次政治和社会动荡的城市，也是一个不断拒绝被归类和定义的城市。

[超胡同]

生长于北京的王硕，现在将工作室设立在旧城的中心，并一直持续关注着胡同的进化和再生。

自 2012 年起，王硕与建筑师安竹 · 布莱恩特（Andrew Bryant）发起并组织了 [超胡同] 项目。[超胡同] 是一项备受殊荣的国际性研究合作项目，它获得格莱汉姆基金会（Graham Foundation）以及麦肯锡城市中国计划（Urban China Initiative）连续 3 年的支持。

[超胡同] 是一个跨学科合作的城市研究平台，旨在调查、揭示和展望正在胡同中展开的城市转变。它由一系列工作坊、研讨会、专题报告会和圆桌会议组成，将胡同视为一种独特的社会文化存在、一个不同城市数据的集合以及城市自我更新的范例。建筑师、城市规划学者、媒体艺术家、社会人类学家和历史学家都将参与到这个活动中，他们的共同目标是通过一种非传统的模型来解译和激发这个世界上最多变和最有活力的城市之一——北京。

当前，北京市中心的胡同俨然已成为这座城市创新改造的实验场地。它们不仅可让人们对北京悠久而带有预示性的历史得以一窥，同时也作为一种文化与物质的存在，为人们提供了一把开启都市未来的钥匙。

当公众舆论正变得越来越沉迷于旧城历史保护的乌托邦的同时，社会发展的压力却使城市经历着快速而大规模的拆除摧毁。在毫不留情的错位现实下，胡同正在以飞快的速度消失着，并且城市化的过程正在趋向于一种困境——悬置于两极之间而没有有效的解决方法。在怀旧的乌托邦和残酷的反乌托邦之间，[超胡同] 关注胡同当下的现实。这一跨学科城市研究的目标是引发讨论并建立一种对胡同这一“城市涌现”的新理解，并借此揭示它所呈现的特质，以及投射未来的种种可能。

Opposite: View from the forecourt toward the West Sea. Photo by Chen Su, ©META-Project.

对页：从前厅看向西海。

Stage 1: Year 2000

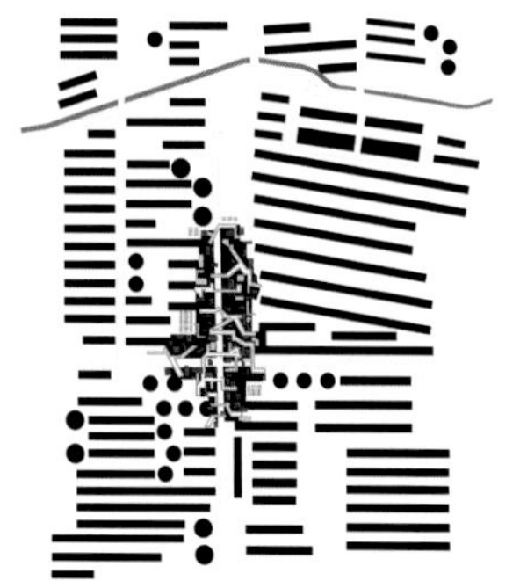

Stage 2: Year 2010

Stage 3: Year 2020

Stage 4: Year 2050

Diagram of program growth in Hutong / 胡同功能的发展图

Fueled by the numerous discussions and knowledge produced from the [META:HUTONGS] research, Courtyard by the West Sea is a parallel attempt on Hutong regeneration that sits in-between preservation and demolition.
Unlike the introverted quality of the traditional courtyard house, the owner of this site asked for a variety of mixed-use programs, including tea house, dining, party space, office, and meeting, as well as dwelling and entertainment. The contemporary and sometime "public" program opened up the courtyard to become "extraverted", so as to induce more human interactions. These required us to break the general understanding of the courtyard as an enclosed typology by introducing the experience of "meandering in the Hutongs" into the courtyard.
We converted the narrow corridor between two rows of brick buildings to a mode compatible with the Hutong-courtyard typology by demolishing the temporary structure to the east and in the middle, so as to introduce cross-sectional changes along the 60-m long site. Then by adding three different types of "loggia" at the hinge of the expanded spaces, we reconstructed a "three-step-courtyard".
Here the "three-step-courtyard" is not an imitation of the traditional symmetrical courtyard pattern in the Hutongs, but a contemporary reinterpretation of the multi-layer courtyard space and its possible variation along the depth, and how it will shift the movement of steps and sense of space. All the mixed-use programs were sorted and divided by three courtyards full of vegetation, making the daily routine of walking in and out of the site a continuous spatial experience full of rhythm.

基于[超胡同]引发的诸多讨论和思考，从研究到平行展开的实践，"西海边的院子"就是META-工作室所进行的关于如何在保护与拆除之间进行改造的一次尝试。

与传统四合院完全"内向性"的居住状态不同，这一宅基地的业主提出了种种混合的功能需求，包括茶室、正餐、聚会、办公、会议以及居住和娱乐。这些现代化的甚至有些"公共性"的功能须在内院里呈现出"外向性"的姿态，从而引发更加开放的人为活动，这促使我们必须打破一般对院落空间围合边界的理解，将近乎于"行走在胡同中"的空间感受引入到院落中来。

META-工作室将形成于两排东西向厂房之间的狭窄、压抑的巷道空间，转化成与胡同院落模式相符的空间类型——选择将东侧破旧的房屋以及南侧厂房的中段拆除，并对一些临时性构筑物进行清理，为贯穿整个60m长的地块中间的狭长走道引入几处剖面宽度上的收放变化。之后，在扩展后的凹凸空间衔接处引入3个不同形式的悬挑门廊，形成空间意义上的"三进院"。

这里所提出的"三进院"，并非是对传统四合院"中轴对称"的院落格局的模仿，而是力图通过错落有致、移步换景的空间层次，以当代的语言重新阐释"多重院落"这一概念在进深变化上的可能，同时构建了业主所期待的胡同文化生活的内涵。三进充满树木植被的院落将业主需要的各种混杂功能合理归纳划分，并使在整个基地内的日常行走成为一种连续而又充满节奏变化的空间体验。

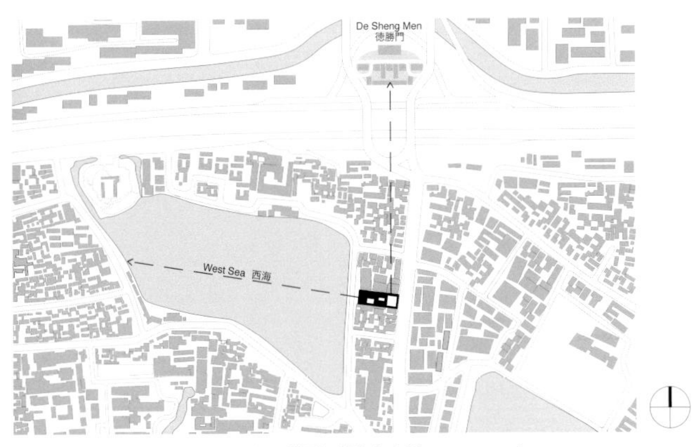

Site plan (scale: 1/10,000) / 总平面图（比例：1/10,000）

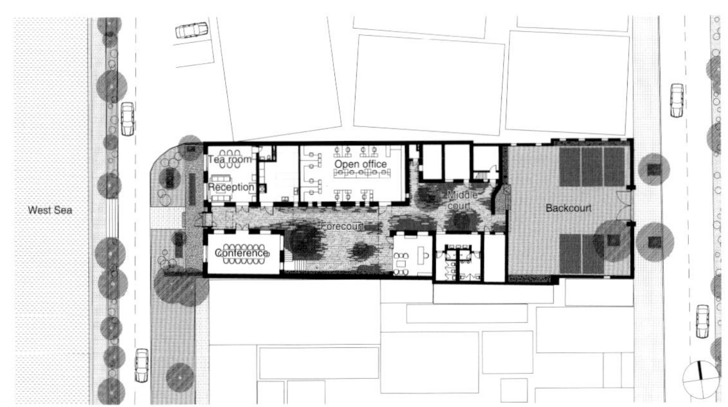

Ground floor plan (scale: 1/1,000) / 首层平面图（比例：1/1,000）

ZAO/standardarchitecture

Micro-Yuan'er

Xicheng District, Beijing, China 2015

标准营造建筑事务所

微杂院

中国，北京市，西城区 2015

This page: View from the southwest. The site is located in Hutong district in Beijing, where many old alleys remain. This project is an intervention to the courtyard of a Dazayuan, which is a Siheyuan (traditional Chinese courtyard residence) that is occupied by many families. Opposite: View from the south. The art space / pavilion is in front, and the library is on the back. Photo by Wang Zilin. All photos on pp. 76–83 except as noted by Su Shengliang / ZAO/standardarchitecture, courtesy of the architects.

本页：从西南侧看场地。基地位于北京的胡同区，该区域保留了许多传统的街巷。该项目是对四合院（多户家庭共同居住的中国传统民居）里大杂院的改造。对页：从南侧看场地。艺术空间/亭子在前面，图书馆在后面。

Credits and Data

Project title: Micro-Yuan'er

Client: Dashilar & Liulichang Culture Development Ltd.

Sponsor: Camerich

Location: Cha'er Hutong #8, Dashilar, Beijing, China

Design: 2012–2015

Architect: ZAO/standardarchitecture

Project architect: Zhang Ke, Zhang Mingming, Fang Shujun

Design team: Zhang Ke, Zhang Mingming, Fang Shujun, Ao Ikegami, Huang Tanyu

Site area: 350 m^2

Building area: 6 m^2 (art space / pavilion), 9 m^2 (library)

建筑与都市
Architecture and Urbanism
Chinese Edition 16:06
064
Feature:
Architects in China
ZAO/standardarchitecture
Micro-Hutong / Micro-Yuan'er
Xicheng District, Beijing, China

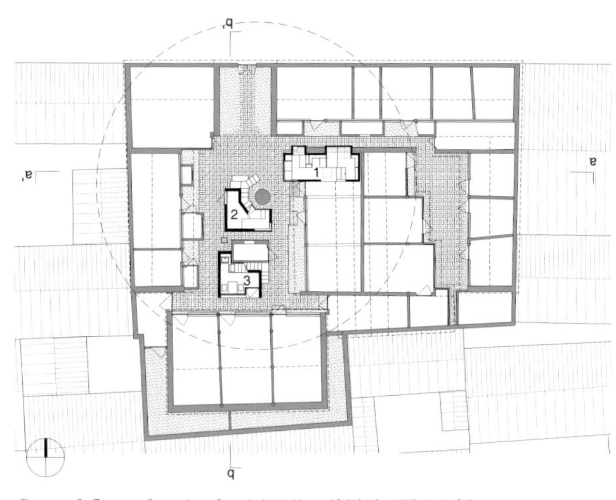

Ground floor plan (scale: 1/500) / 首层平面图（比例：1/500）

Cha'er Hutong (Hutong of tea) is a quiet spot in the busy Dashilar area, situated 1 km from Tiananmen Square in the city centre. No. 8 Cha'er Hutong is a typical "Da-Za-Yuan" (big-messy-courtyard) once occupied by over a dozen families. Over the past five decades each family built a small add-on kitchen in the courtyard. These add-on structures form a special density that is usually considered urban scrap and almost all of them have been automatically wiped out with the renovation practices of the past years.

In this project the architects tried to redesign, renovate and re-use the informal add-on structures instead of eliminating them. In doing so, they intend to recognize the add-on structures as an important historical layer and as a critical embodiment of Beijing's contemporary civil life in Hutongs that has so often been overlooked.

In symbiosis with the families who still live in the courtyard, a 9-m² children's library built of concrete mixed with Chinese ink was inserted underneath the pitched roof of an existing building. Under a big ash tree, one of the former kitchens was redesigned into a 6-m² mini art space made from traditional bluish grey brick. On its exterior, a trail of brick stairs leads up to the roof, where one may delve into the branches and foliage of the ash tree. With the small-scale intervention in the Cha'er Hutong courtyard, the architects try to strengthen bonds between communities, as well as to enrich the Hutong life of local residents. A child may stop by after school, pick out a favorite book, and read in his little niche before getting picked up by the parents. Or the kids may climb up onto the roof, sit in the shade, and engage in a cozy conversation with the elderly in a familiar but new space.

茶儿胡同 8 号院是热闹繁忙的北京大栅栏地区的一片“静土”，距离市中心的天安门广场仅千米远。它是一个典型的“大杂院”，一度有十几户人家居住在这里。在过去的 50 年里，每家每户都在院子里加建了小厨房。这些附加结构形成了一种特殊的密度，它们通常被认为是城市废弃物，在以往的改造实践中几乎全部被毫不犹豫地予以清除。

在该项目中，建筑师试图对这些非正式的附加结构进行重新设计、改造和利用，而不是一味地消除它们。他们希望通过这种做法来为附加结构正名，使人们意识到它们是历史进程中的重要一环，更是经常被人们忽略的当代北京胡同生活的具体体现。

建筑师在现有建筑的斜屋顶下植入了一个 9 m² 的儿童图书馆，并用胶合板来建造，以使其与仍居住在四合院中的家庭和谐共生。此外，建筑师还将原来高大的国槐树下的厨房重新设计为一个以传统灰砖建造的 6 m² 的迷你艺术空间。建筑外表面的砖砌楼梯直通屋顶，可以近距离地观察槐树的枝叶。建筑师试图通过小尺度的茶儿胡同四合院来增加社区间的联系，同时丰富当地居民的胡同生活。孩子们放学后可以挑选一本自己最喜爱的书，在等待父母来接自己的时候坐在壁龛里阅读；或和小伙伴们一起爬到屋顶，坐在树荫下，在既熟悉又新奇的空间里与老人们随意地聊天。

This page: View of library for children. Photo by Zhang Mingming / ZAO/ standardarchitecture. Opposite: View from the top of the steps surrounding the ash tree. An existing gate leading from the street to the courtyard is on the north of the site.

本页：看向儿童图书馆。对页：从环抱着国槐树的台阶上往下看。基地北侧是一扇原有的由大街通向中庭的门。

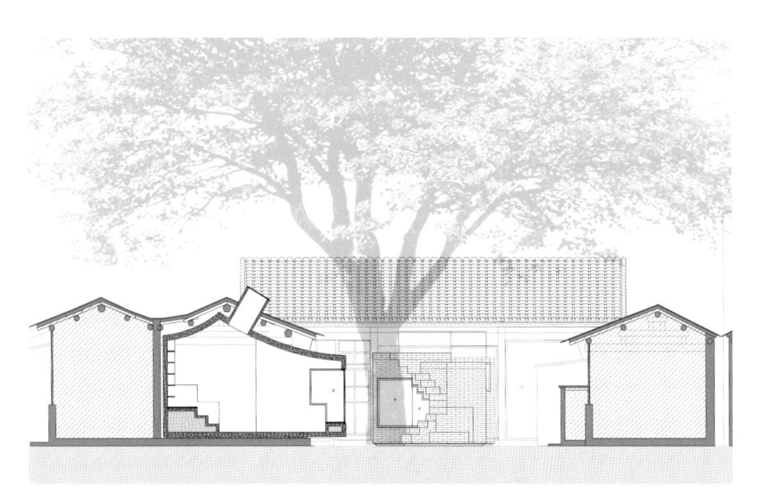

a-a' section (scale: 1/200) / a-a' 剖面图（比例：1/200）

b-b' section / b-b' 剖面图

ZAO/standardarchitecture
Micro-Hutong
Xicheng District, Beijing, China 2013

标准营造建筑事务所
微胡同
中国，北京市，西城区 2013

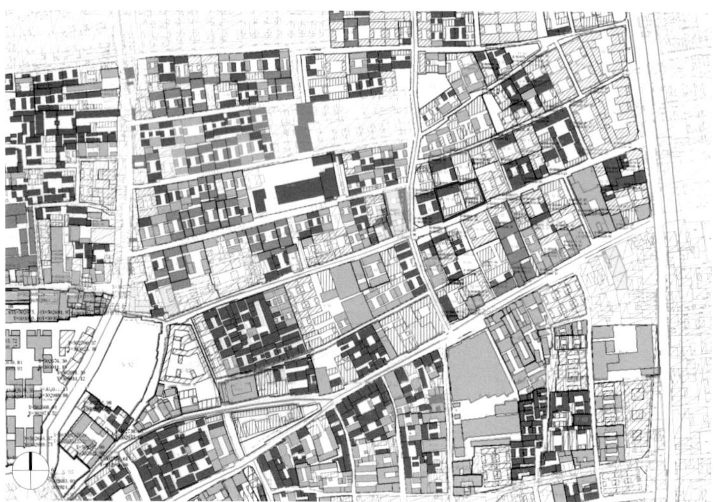

Site plan (scale: 1/15,000) / 总平面图（比例：1/15,000）

p. 80: The project uses the traditional small scale and courtyard of the Hutong. Photo by Chen Su. p. 81: Looking up from the courtyard. This page: Exterior view from the street. The entrance leads to the courtyard. Opposite: General view looking down from the south. Photo by Chen Su.

80页：项目利用了胡同传统的小尺度和庭院。81页：从庭院往上看。本页：从街道上看项目的外观。引人入院的入口。对页：从南侧往下看的全景。

Micro-Hutong is a building experiment by Zhang Ke's ZAO/standardarchitecture team on the Yang-Mei-Zhu street of the Dashilar area. The goal of the project is to search for possibilities o creating ultra-small-scale social housing within the limitations of super-tight traditional Hutong spaces of Beijing.

Located in the Dashilar District, a historical area within walking distance of Tiananmen Square, ZAO/standardarchitecture has designed a 30-m² Micro-Hutong that offers a new alternative to Hutong preservation and actualization.

A critical look into the dynamics of the Hutong reveals that even with the menacing grip of unscrupulous real estate development, the most critical problem of the Hutong consists of the relentless exodus of its occupants. Concerned with the lack of facilities and the absence of quality communal space, they decide to sell and move out to bigger apartments outside of the city centre. This constant desertion of the traditional dweller of the Hutong from the heart of Beijing prompted the generation of a strategy able to challenge the growing disinterest of the Hutong tenant in order to keep alive valuable living traditions.

The result is an architectural operation that brings back the courtyard as a generator of program, as it activates the building by creating a direct relationship with its urban context, drawing to its interior social activities. Apart from enhancing the flow of air and light, the courtyard creates a direct relationship between the living space contained in the dynamic volumes and an urban vestibule in the front part of the building. This flexible urban living room acts as a transition zone from the private rooms to the street, while serving as a semi-public space to be used by both the inhabitants o the Micro-Hutong and the neighbors of the community.

The Micro-Hutong inherits the intimate scale of the traditional Hutong, revitalizing its social condensing capabilities, while enhancing it with spatial improvements. Its light-steel structure and plywood panel cladding allows for low-cost construction, while creating new possible reconfigurations for the future of the Hutong in Beijing.

微胡同是张轲领导的标准营造建筑事务所在北京大栅栏杨梅竹斜街进行的一次建造实验，目的是探索在传统胡同局限的空间中创造可供多人居住的超小型社会住宅的可能性。

微胡同位于北京大栅栏历史文化街区内，距离天安门广场很近，步行可达。标准营造设计的这座约 30m² 的微胡同试图为胡同保护与更新提供一种新方式。

基于现状而批判性地思考，对胡同文化最大的威胁不是恣意延展的商业开发，而是生长于此、满载着历史记忆的大批原居民的离去。由于缺乏较好的基础生活设施和有品质的公共空间，居民们决定租售这些房子，搬迁至市郊更宽敞的居所里。老住户被迫遗弃胡同的现象持续上演，亟需一种应对策略以留住这里曾有的充满生机的居住传统。

设计将庭院重新作为流线组织的重心，通过将活动空间引入建筑内部庭院，来创造与城市文脉的直接联系。庭院不仅提升了内部空气与光线的流动，也连接着形式多样的方形结构及面向城市的门廊。这个灵活的城市居住空间成为较私密的生活空间与具有城市性的街道间的过渡空间，同时也成为可供微胡同居民及社区邻居共同使用的半公共空间。

微胡同继承了传统胡同所具有的亲密空间，也复兴了其社会功能，同时增强了它的空间特性。它使用的轻钢结构及胶合板面材，保证了其低造价的实施，可发展成为北京胡同更新保护的可行性范本。

Credits and Data

Project title: Micro-Hutong
Client: Dashilar & Liulichang Culture Development Ltd.
Sponsor: Camerich
Location: YangMeiZhu Xie street, Dashilar, Beijing, China
Design phase: March 2012 – September 2013
Construction phase: September 7, 2013 – September 25, 2013
Architect: ZAO/standardarchitecture
Project architect: Zhang Ke, Zhang Mingming
Design team: Zhang Ke, Zhang Mingming, Dai Haifei, Ao Ikegami, Zhang Yanping, Huang Tanyu, Zhang Zhaosong
Site area: 60 m²
Building area: 48 m²

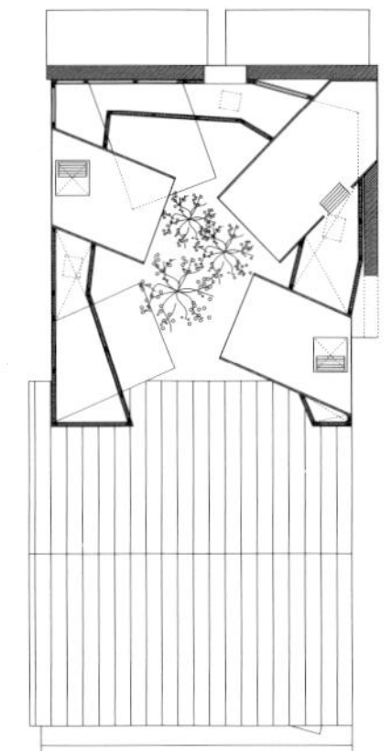

Second floor plan / 二层平面图

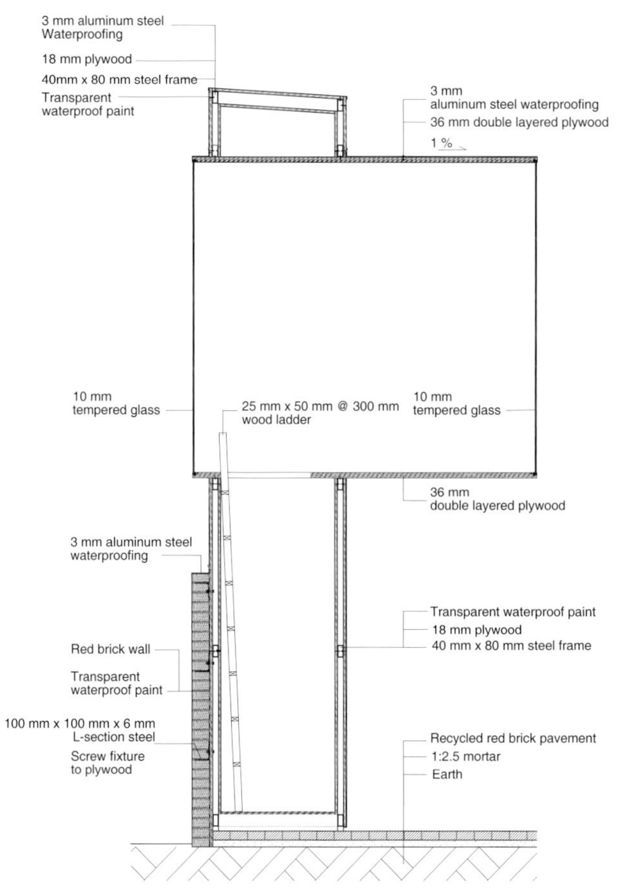

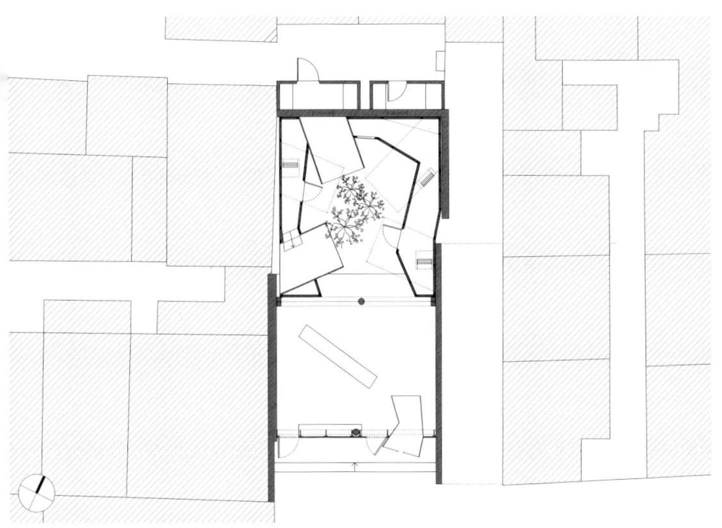

Ground floor plan (scale: 1/300) / 首层平面图（比例：1/300）

Section detail of wall (scale: 1/50) / 墙体剖面详图（比例：1/50）

Chen Haoru

Bamboo Design in the Sun Farming Commune

Hangzhou, Zhejiang, China 2014

陈浩如

太阳公社竹构系列

中国，浙江省，杭州市 2014

Credits and Data

Project title: Bamboo Design in the Sun Farming Commune

Client: Taiyang Organic Farming Commune

Program: Pig barn, henhouse, pavilion

Location: Shuangmiao Village, Lin'an Township, Hangzhou, China

Design: 2013

Completion: 2014

Architects: Chen Haoru, Xie Chenyun, Ma Chenglong, Wang Chunwei

Materials: Bamboo, thatch, stone

Pig barn site area: 2,000 m²

Pig barn built area: 256 m²

Henhouse built area: 130 m²

Pavilion built area: 120 m²

Chen Haoru
Bamboo Design in the Sun Farming Commune / Hangzhou, Zhejiang, China

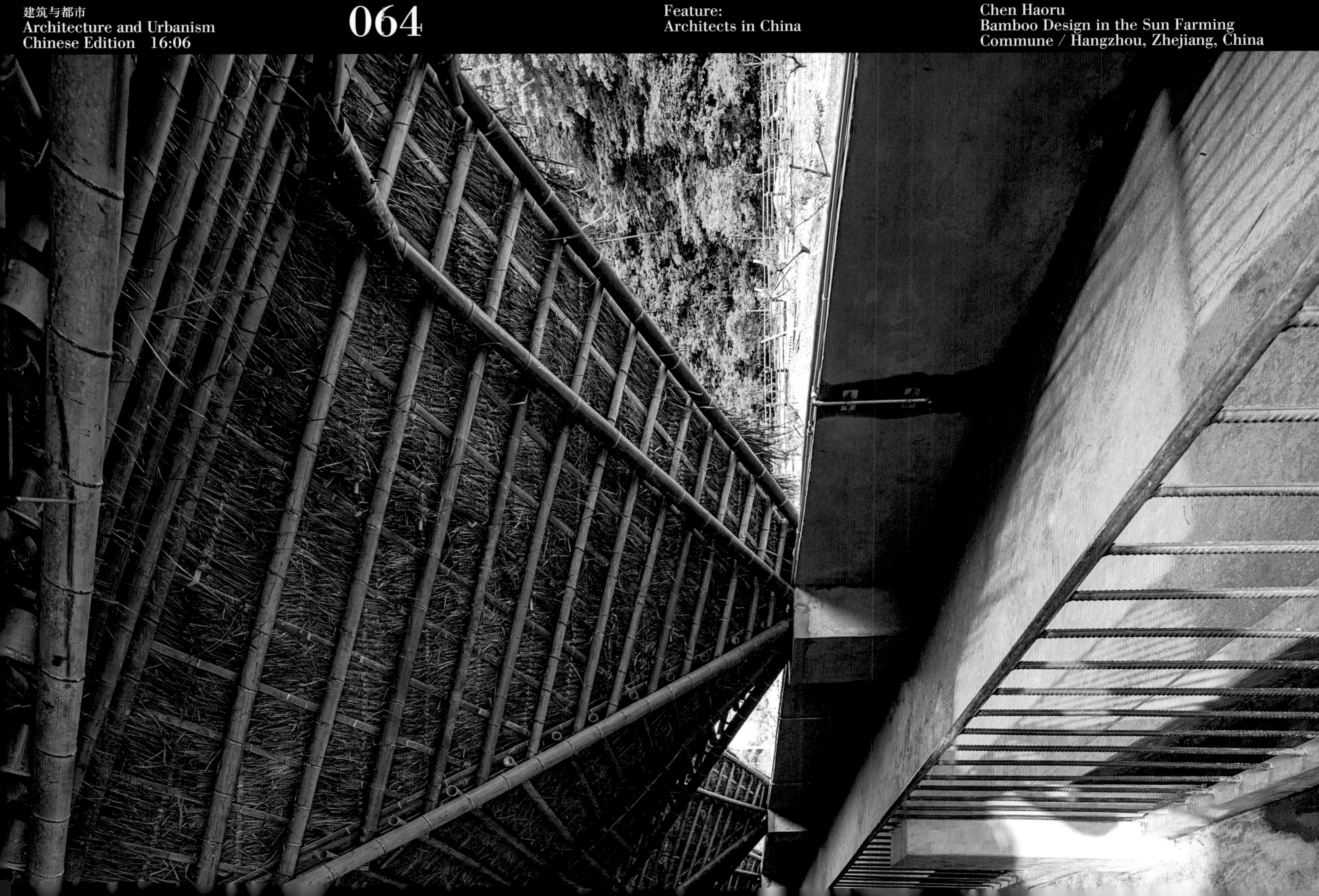

This project base, located in the mountainous area west of Hangzhou, is a natural village with 140 peasant households. The main building materials are bamboo that grows in its original site, and pebbles picked from the rivulet. The craftsman named Chen is the eldest son of a farmer in this area, other workers and materials are all local. During the design process, animal behavior was studied while rotational grazing and site were planned at the same time.

The pig barn is designed with a special dormitory, feeding area, outdoor toilet, external field and swimming pool. Excavation is no longer needed on the field. A self-stable structure which seems like a big bird landed on the stone parapets, was completed using giant bamboo. Thatch picked in a mountainous area nearby was first knitted by villagers at the commune's call, and then hung onto the roof of this bamboo structure by the workers.

The henhouse is built on the compacted flat ground between the north hilly area and the small agricultural reservoirs at the end of the valley. The shallow foundation is formed by uniformly inserting tiny wood elements into the ground. With the use of bamboo, a platform is built on top of the wood foundation, on which are placed two 8×8 m bamboo unit structures to form an open and sturdy structure. A quickly and easily built bamboo roof creates stability for the whole construction. The interior is tall and roomy, and covered densely with bars, which allows the birds stay on them.

The pavilion was built on the reservoir dam with a great view and good ventilation. The foundation of the pavilion is the top of the dam. It is consolidated to set the base with six logs and connect to the plank covered above. With the bottom timber, five bays of bamboo structure form an independent and compact structure. The unit for one bay is 4×3 m to build this pavilion suitable for people to live, and the function therefore is to be a resting place for farmers in work.

This project is a social experiment on natural construction with environmental design. Based on the reflection of mess made by urban construction, we tried to make a difference in several aspects. With no outside intervention and using only local labor and materials to set up all natural structures is thought to be the first consideration of the concept. Harboring the idea of the signification in rural construction, we activate the village by designing this project of eco-farm, and building up a new image and new economic system. This is a positive social experiment for the architect, to be involved and to involve all rural force into one shared objective.

项目基地位于杭州西部的山区，该地区是一个拥有 140 户农户的自然村落。猪圈主要由原场地生长的毛竹构建和溪坑卵石砌筑，工匠陈师傅是场地农民的大儿子，其他工人和材料也均来自当地。设计过程中对动物的习性作了研究，并将轮牧方式和饲养场地作了规划。

猪圈有专门的宿舍、喂食区、外置厕所、室外活动场和泳池。场地上不再动土开挖，巨型青竹结构构成一个自我稳定的结构体，如同一只大鸟落在石砌的矮墙上。手工编织的茅草采自附近山区，由公社发动村民集体编制后，再由工人干挂到竹构屋顶上。

鸡舍基地位于山谷尽端的农业小水库与北侧山地之间夯实的平地上，由小细木均匀地插入地面形成浅基础。木基础的顶部以毛竹搭出一个平台，两个 8m×8m 的竹构单元置于其上，形成一个开敞而结实的结构。屋顶为可快速搭建的简单竹瓦构造，对结构起到稳定作用。竹构内部高挑宽敞且杆件密布，可容纳禽类栖居。

长亭建在水库堤坝上，视野开阔，通风良好。其基础为早已夯实的堤坝顶部，在其上放置 6 根原木作为基础，上铺木板相连，五开间的竹构和底部的原木及木板形成一个独立而紧密的结构体。每个单元的开间为 4m×3m，尺度适合人居，功能为农民在工作时休憩使用。

“太阳公社”是具有社会意义的项目。在对城市建筑灾难进行反思后，建筑师尝试了一种在乡村环境下的自然构造。该实验涵盖多个方面，首先就是在基本没有外力协助的前提下，使用当地的材料及人工，建筑师在
完成建筑构造。以竹构和茅草为主
系列自然构造适合田间地头的乡村
施，易于重修。这是一种当地的手
艺传统重建项目，也是乡村生活的
造，旨在通过“在地建造”激活乡村
协助乡村重建形象和经济体系。

pp. 84–85: Interior view of the pig barn. Photo by Lv Hengzhong. This page: Distant view of the henhouse. The roof is built with bamboo tiles. Photo courtesy of the architect. Opposite, above: Night view of the pig barn. The roof is thatched. Photo by Lv Hengzhong. Opposite, below left: Sketch of the pavilion. Opposite, below right: Bamboo roof structure of the pig barn. Photo courtesy of the architect.

84–85页：猪圈内景。本页：遥望鸡舍。屋顶是用竹片做成的。对页，上：猪圈夜景。屋顶是用茅草覆盖的。对页，左下：亭子的草图。对页，右下：猪圈的竹制屋顶结构。

Site plan (scale: 1/25,000) / 总平面图（比例：1/25,000）

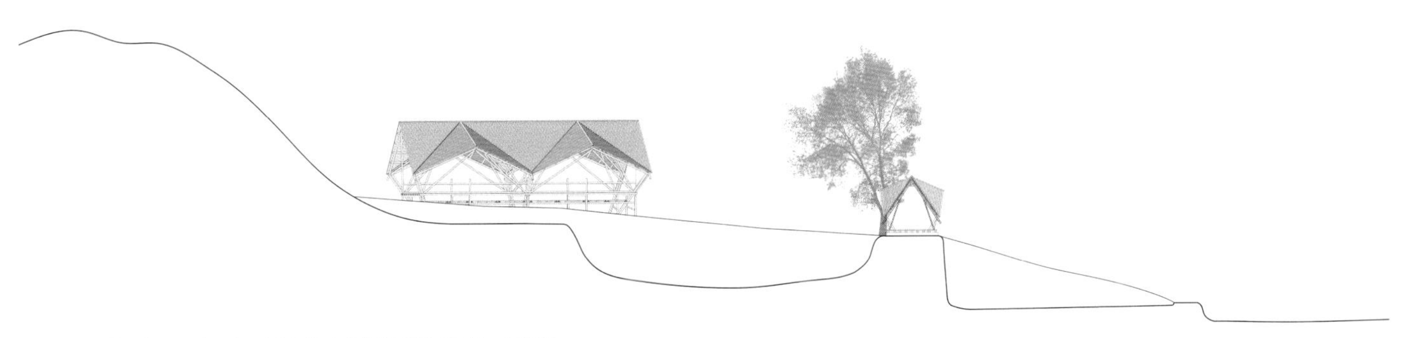

Section of henhouse (scale: 1/500) / 鸡舍剖面图（比例：1/500）

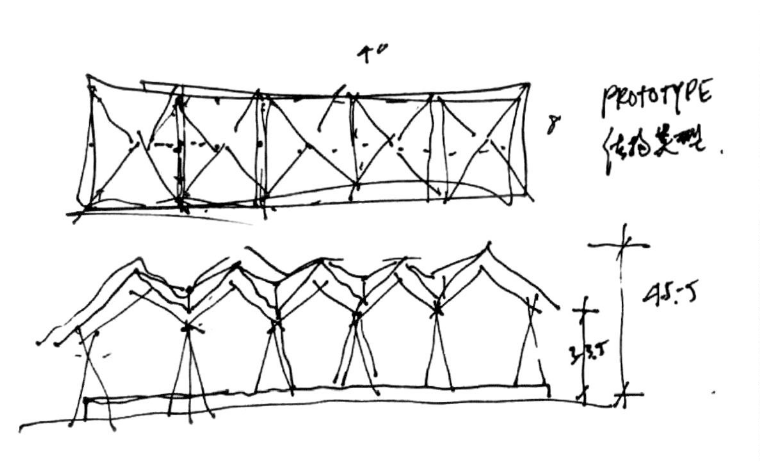

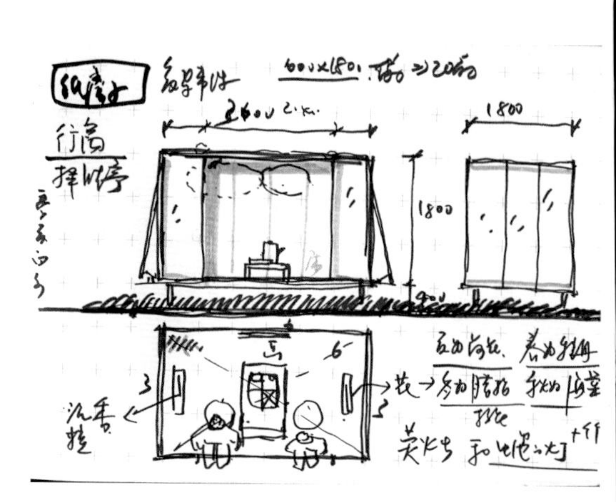
1800
1800
沉香
花

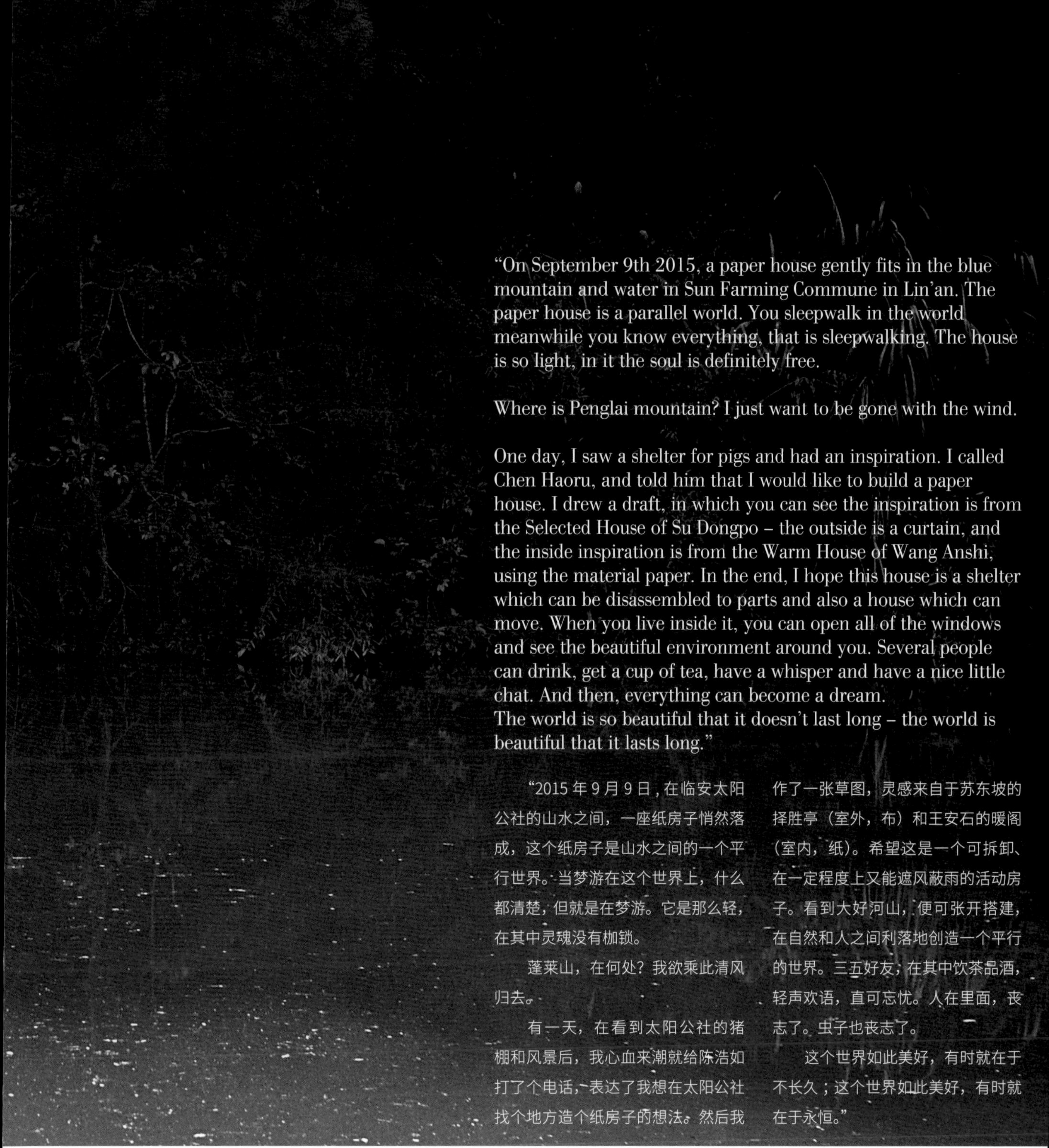

"On September 9th 2015, a paper house gently fits in the blue mountain and water in Sun Farming Commune in Lin'an. The paper house is a parallel world. You sleepwalk in the world meanwhile you know everything, that is sleepwalking. The house is so light, in it the soul is definitely free.

Where is Penglai mountain? I just want to be gone with the wind.

One day, I saw a shelter for pigs and had an inspiration. I called Chen Haoru, and told him that I would like to build a paper house. I drew a draft, in which you can see the inspiration is from the Selected House of Su Dongpo – the outside is a curtain, and the inside inspiration is from the Warm House of Wang Anshi, using the material paper. In the end, I hope this house is a shelter which can be disassembled to parts and also a house which can move. When you live inside it, you can open all of the windows and see the beautiful environment around you. Several people can drink, get a cup of tea, have a whisper and have a nice little chat. And then, everything can become a dream.
The world is so beautiful that it doesn't last long – the world is beautiful that it lasts long."

"2015 年 9 月 9 日，在临安太阳公社的山水之间，一座纸房子悄然落成，这个纸房子是山水之间的一个平行世界。当梦游在这个世界上，什么都清楚，但就是在梦游。它是那么轻，在其中灵魂没有枷锁。

蓬莱山，在何处？我欲乘此清风归去。

有一天，在看到太阳公社的猪棚和风景后，我心血来潮就给陈浩如打了个电话，表达了我想在太阳公社找个地方造个纸房子的想法。然后我作了一张草图，灵感来自于苏东坡的择胜亭（室外，布）和王安石的暖阁（室内，纸）。希望这是一个可拆卸、在一定程度上又能遮风蔽雨的活动房子。看到大好河山，便可张开搭建，在自然和人之间利落地创造一个平行的世界。三五好友，在其中饮茶品酒，轻声欢语，直可忘忧。人在里面，丧志了。虫子也丧志了。

这个世界如此美好，有时就在于不长久；这个世界如此美好，有时就在于永恒。"

Credits and Data
Project title: Bamboo House
Client: Yu Ting, Sun commune
Location: Sun commune, Lin'an, Zhejiang, China
Design time: August 31, 2015–September 6, 2015
Architects: Yu Ting, Zhu Chen
Design architect: Yu Ting
Director: Zhu Chen
Project team: Yu Ting, Zhu Chen, Zhang Jingwen
Associates: CITIARC, Sun commune bamboo craftsman
Floors: 1
Building height: 1.8 m
Structure: Bamboo structure
Main materials: Bamboo and paper
Site area: 7 m^2
Building area: 7 m^2

pp. 88–89: General view. Opposite, below left: Close-up of the paper wall. Opposite, below middle: Close-up of the bamboo structure. Opposite, below right: Sketch by the architect. All photos on pp. 88–89 by Chen Jie, courtesy of the architect.

88–89页：全景。对页，左下：纸墙的近景。对页，中下：竹制结构的近景。对页，右下：建筑师的手绘草图。

Atelier Li Xinggang

Jixi Museum

Xuancheng, Anhui, China 2013

李兴钢工作室

绩溪博物馆

中国，安徽省，宣城市 2013

Jixi Museum is located on the north side of the old town area in Jixi County, Anhui Province. The site was the seat of the town hall throughout history and different regimes. Now because of the comprehensive preservation and restoration program for the entire old town, the site is reprogramed and a comprehensive museum on local culture and history that will have exhibition spaces, a 4D cinema, citizen services, shops, administration and storage functions will be located here.

The design is based on surveys of the landscape and origins of Jixi's name, and investigations and studies on the Hui-style (the style specific to the district named Huizhou during ancient times) settlements. The entire building is covered by a continuous roof with an undulating profile and texture that mimics the mountains and waters surrounding the county. It is the logical result and expression of "Jixi's shape". Once the other buildings in the old town are restored back to the Hui style, the museum will fit in even more naturally with the entire town.

In order to preserve as many of the existing trees on the site as possible, multiple courtyards, patios and lanes are introduced into the overall layout of the building. At the same time, the layout is as a reinterpretation of the spatial layout of vernacular Huizhou architecture, and creates comfortable outdoor spaces. Two streams run along the lanes, one on the east and the other on the west of the cluster of buildings, finally converging into the pool in the large courtyard at the main entrance. In the south part of the building is "Mingtang", an interior courtyard common in the typical layout of Hui-style houses, following the traditional Chinese Fengshui principles. Directly opposite the main entrance is a group of abstract "rocks". Surrounding "Mingtang" is a "sightseeing route" that guides tourists to the "sightseeing platform" at the southeast corner of the building, where they can have a bird's eye view of the roofscape, courtyards and distant mountains.

Triangular steel structural trusses (their slopes are derived from the local building) adapt well to the undulating roof. Local building materials such as stone, clay roof tiles are used in modern and innovative ways to pay respect to history yet respond to our own times.

Credits and Data

Project title: Jixi Museum

Location: Jixi, Anhui, China

Design period: November 2009–December 2010

Construction period: December 2010–November 2013

Team: Li Xinggang, Zhang Yinxuan, Zhang Zhe, Xing Di, Zhang Yiting, Yi Lingjie, Zhong Manlin

Structure: Wang Libo, Yang Wei, Liang Wei

Landscape: Li Li, Yu Chao

Site area: 9,500 m²

Building area: 10,003 m²

绩溪博物馆位于安徽省绩溪县的旧城北部，基地曾为县衙，后建为县政府大院，现因古城整体被纳入保护修整规划，所以改变原有功能，将其改建为博物馆。项目包括展示空间、4D 影院、观众服务、商铺、行政管理和库藏等功能空间，是一座中小型的地方历史文化综合博物馆。

设计基于对绩溪的地形环境、名称由来的考察和对徽派建筑与聚落的调查研究。整个建筑被覆盖在一个连续的屋面下，起伏的屋面轮廓和肌理仿佛绩溪周边的山形水系，是"北有乳溪，与徽溪相去一里，并流离而复合，有如绩焉"中"绩溪之形"的充分演绎和展现。待周边区域修整"改徽"完成，古城风貌恢复后，建筑将与整个城市形态更自然地融为一体。

为尽可能地保留用地内的原有树木（特别是用地西北部的一株有 700 年树龄的古槐），设计在建筑的整体布局中设置了多个庭院、天井和街巷，既营造出舒适宜人的室内外空间环境，也是对徽派建筑空间布局的重释。建筑群落内，沿着街巷设置有东西两条水圳，汇聚于主入口大庭院内的水池。在建筑南侧设置内向型的前广场——"明堂"，符合徽派民居的典型布局特征，同时也符合中国传统中"聚拢风水之气"的理念。主入口正对方位设置一组被抽象化的"假山"。围绕"明堂"、大门和水面，有一条对市民开放的立体的"观赏流线"，可将游客引至建筑东南角的"观景台"，俯瞰建筑的屋面、庭院和秀美的远山。

规律性组合布置的三角形屋架单元的坡度源自当地建筑，并适应连续起伏的屋面形态。建筑在适当采用当地传统建筑技术的同时，也灵活地使用石、瓦等当地常见的建筑材料，并尝试使之呈现出当代感。

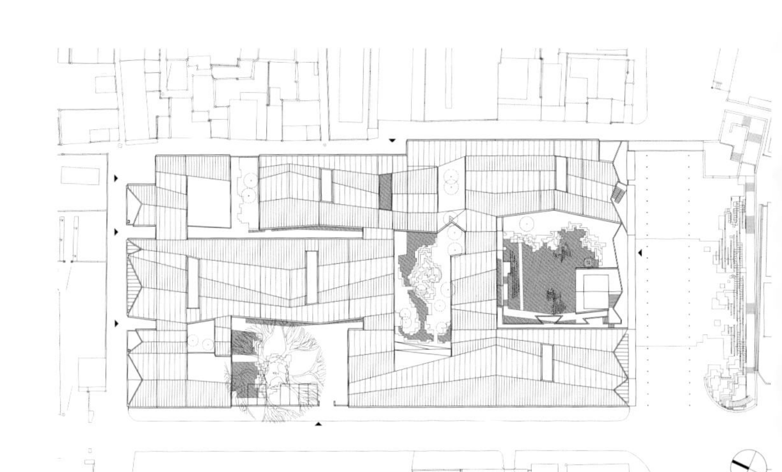

Site plan (scale: 1/2,000) / 总平面图（比例：1/2,000）

Atelier Li Xinggang
Jixi Museum
Xuancheng, Anhui, China

p. 91: Aerial view from the southwest. Photo by Li Zhe. pp. 92–93: View of the continuous roofs from the south. The undulating roofs refer to the texture of the surrounding landscape. Photo by Qiu Jianbing. All photos on pp. 91–95 courtesy of the architect.

91页：从西南侧俯瞰场地。92–93页：从南侧看连绵的屋顶。起伏的屋顶是对周边环境质感的回应。

Courtyard
Preface hall
Reception hall
VIP room
. Classroom
. Shop
. Ticket
. Tea pavilion
. Reserved relic site
. Exhibition hall
. 4D cinema
. Temporary exhibition hall
. Lecture hall
. Equipments
. Fire control room
. Technology and management room
. Temporary storage
. Storage facilities
. Lane
. Preparation room
. Lavatory

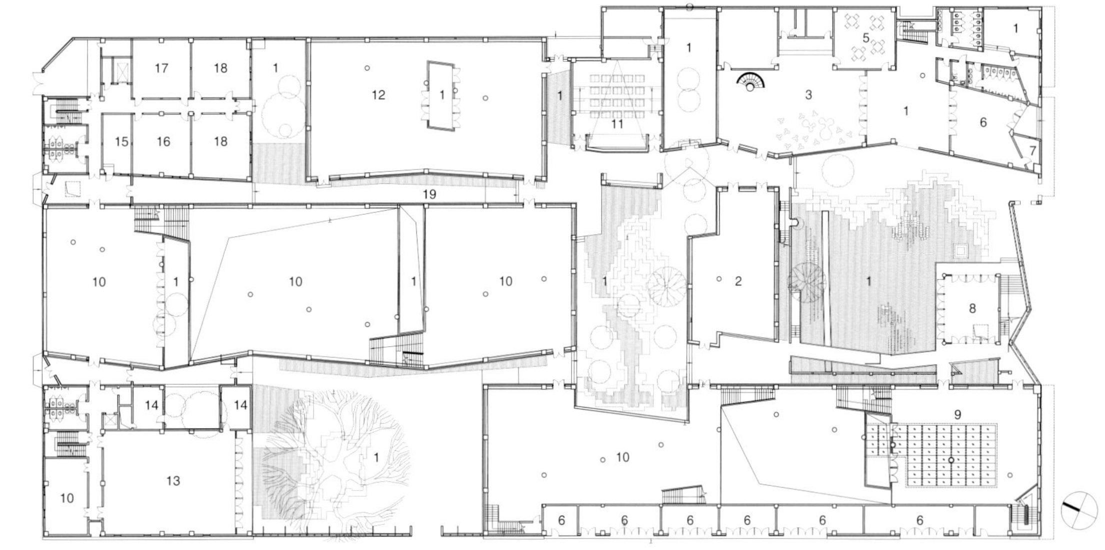

Ground floor plan (scale: 1/1,000) / 首层平面图（比例：1/1,000）

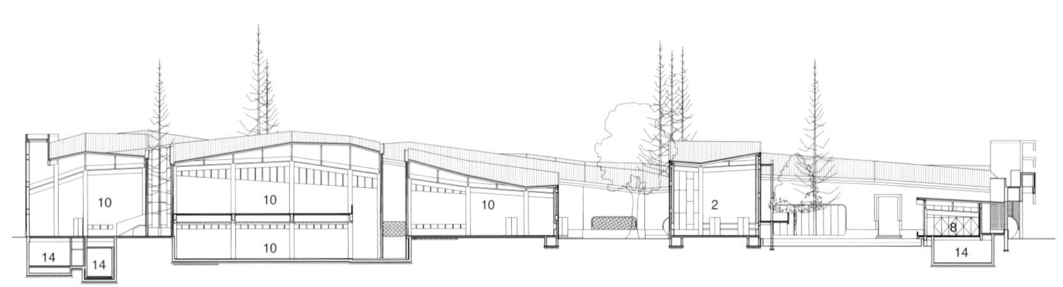

Longitudinal section (scale: 1/1,000) / 纵向剖面图（比例：1/1,000）

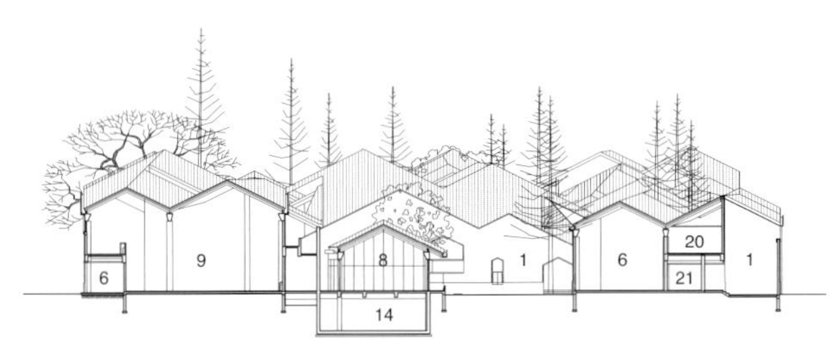

Cross section / 横向剖面图

Opposite, above: View of the courtyard with an old Chinese scholar tree. Opposite, below: The lane with stream ditch connecting to the north entrance comes from the lane and drainage system of ancient Huizhou villages. This page, left: Aerial view of the path around the courtyard next to the main entrance, which provides an experience of a traditional Chinese garden. This page, right: Interior view of the exhibition hall on the southwest side. All photos on pp. 94–95 by Xia Zhi.

对页，上：有一棵老槐树的院子。对页，下：连接北入口的有水圳的小道是古老徽州村落小道和排水系统的延续。本页，左：俯瞰主入口附近环绕中庭的小路，它营造出一种中国古典庭园的空间体验。本页，右：西南侧展览厅的内景。

TeamMinus

Janamani Visitor Centre

Yushu, Qinghai, China 2012

简盟工作室

嘉那嘛呢游客到访中心

中国，青海省，玉树市 2012

TeamMinus
Janamani Visitor Centre
Yushu, Qinghai, China

Credits and Data
Project title: Jianamani Vistor Centre
Client: Jianamani Tourism Administration
Program: Visitor centre
Location: Yushu, Qinghai, China
Period: 2011–2012
Architects: TeamMinus
Materials: Stone, timber, steel
Area: 1,121 m²

pp. 96–97: Distant view from the south. Janamani, Tibetan Buddhist cairn is on the left. pp. 98–99: Exterior view of northeast facade. All photos on pp. 96–101 by BU Lei, courtesy of the architect.

96–97页：从南侧遥望场地。左侧是藏传佛教的嘉那嘛呢石经城。98–99页：建筑东北立面的外观。

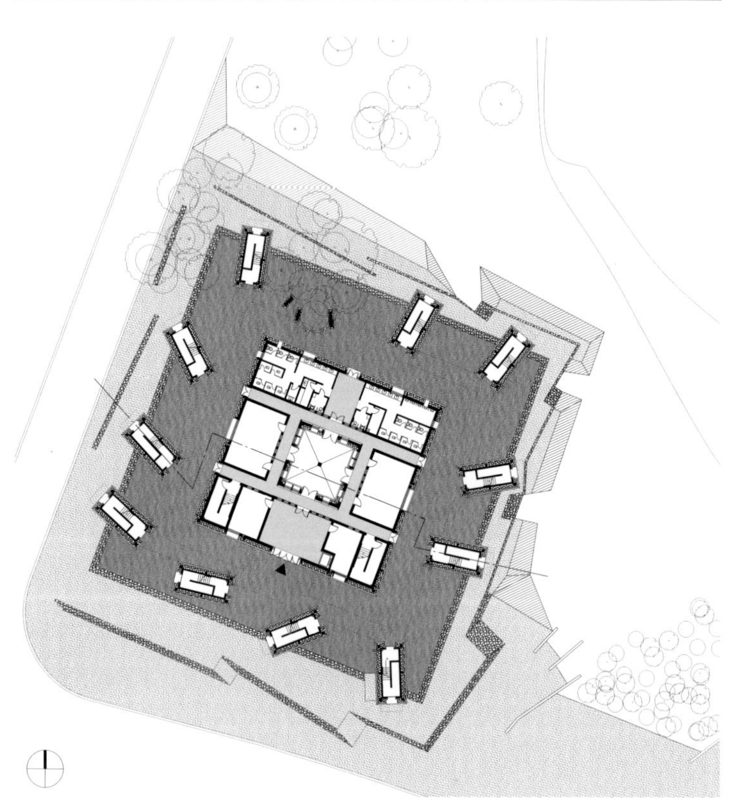

Ground floor plan (scale: 1/1,000) / 首层平面图（比例：1/1,000）

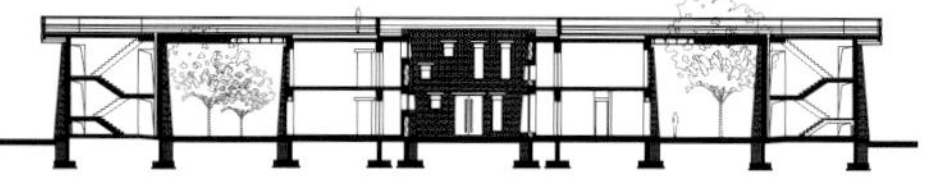

Section (scale: 1/1,000) / 剖面图（比例：1/1,000）

Yushu is a highly regarded religious centre to Tibetans. Its significance comes mainly from Janamani, the world's largest Tibetan Buddhist cairn. With a history of over three centuries, Janamani currently bears over 250 million pieces of Mani stone (stones inscribed with sutras or mantras), and is still growing with new pieces added daily by pilgrims. In Yushu, more than 40% of the population live on the carving of Mani stones. To the Yushu community, nothing compares to Janamani. After the 2010 earthquake, Yushu-ers immediately set off to repair Janamani, long before they started repairing their own houses.
The Janamani Visitor Centre serves both visitors and the local community. To visitors and pilgrims, it provides information about Janamani and its history complemented by views of the real sites. To local rresidents, it provides a post office, a clinic, public toilets and a small research archive.
The Janamani Visitor Centre consists of a square building with courtyard in the centre, and 11 observation decks surrounding it. The central square volume features the typical Tibetan layout. Of the 11 observation decks, two point to Janamani, and nine points to historic / religious sites related to Janamani, including: Leciga, Genixibawangxiou, Cuochike, Dongna Zhunatalang Taiqinleng, Zhaqu River Valley, Lazanglongba, Rusongongbu, Naigu River Beach, and Guanyin Rebirth Site.
The Janamani Visitor Centre is mainly built by local construction techniques. The stone masonry is done by local masons, using the same kind of local rock from which Mani stones are carved. The railings around the roof terrace and the observation decks are made of wood, with some parts recycled from earthquake debris.

玉树市是藏区极其重要的宗教中心之一，这主要是源自新寨的嘉那嘛呢石堆。它是由各方信徒堆放而成的，历经 300 余年，目前约有 2.5 亿块嘛呢石，规模为世界之最。玉树近 40% 的人口以雕刻嘛呢石为生。嘉那嘛呢石堆在玉树人的心中占据着无可比拟的地位。在地震之后，玉树人在修复自家住宅之前首先修复的是嘛呢石堆。

嘉那嘛呢游客到访中心，一方面通过展示空间与观景台对嘉那嘛呢的历史文化进行价值诠释，服务到访的信徒与游客；另一方面，集成邮局、诊所、嘛呢石研究机构及公共卫生间等，服务玉树本地社区。

嘉那嘛呢游客到访中心由“回”字形功能空间和环绕其周边的 11 个观景台组成。中央的四方空间设计具有典型的藏族特点。观景台中有 2 个指向嘛呢石堆，其余 9 个分别指向勒茨噶、格尼西巴旺秀、错尺克、洞那珠乃塔郎太钦楞、扎曲河谷（通天河）、拉藏龙巴、茹桑贡布神山、乃古滩、观世音轮回道场等 9 处嘉那嘛呢宗教活动的圣地或嘉那嘛呢历史上的重要地点。

嘉那嘛呢游客到访中心是使用地方材料和工法建造的。砌筑的石头与雕刻嘛呢石的石材相同，并由当地的石匠打磨加工。屋顶露台和观景台的扶手均为木制，其中的部分木材是回拣再利用的震后房屋残骸。

This page, above: View of pilgrimage. Mani stones left by pilgrims are amassed. This page, below: View of the pedestrian connecting between the roof terraces. Opposite, above: View of the central courtyard. Opposite, below: Exterior wall of the ground floor is made of the same local rock as Mani stone.

本页，上：朝圣的景象。由朝圣者堆积而成的嘛呢石堆。本页，下：屋顶露台之间的人行通道。对页，上：中央庭院。对页，下：一层的外墙由和嘛呢石堆相同的本土石头建成。

Zhaoyang Architects

Artist's House in Xizhou

Dali, Yunnan, China 2016

赵扬建筑工作室

喜洲画家住宅

中国，云南省，大理市 2016

This is a house designed for a painter and his wife. The site is located on the eastern edge of a village, next to rice fields. In common with the inward-looking character of courtyard houses in the village, the new house has an introverted character. Lime mixed with pieces of straw – a very common form of rendering for external walls in the Dali region – also helps relate the new house to its context.

Inside, the house is divided into the private quarters of the resident couple to the north, and a series of living areas and accommodation for a guest and a live-in cleaner / cook to the south. Nine courtyards of varying sizes reduce the scale of the traditional courtyard house and match courtyards with various rooms. The circulation through the main parts of the house develops as an alternation between spatial compression and spatial dilation. The entry sequence in the southwest involves two 180-degree turns that bring the visitor to the southern courtyard. A long corridor offers quick access to the private quarters, with a glimpse of the central shallow pond halfway. From the southern courtyard one can walk under cover past an enclosed dining area towards the pond. A storage room obscures the size of the main living area, with views out to the eastern fields as well as to the central pond and, above the plants and rocks of the western passage, to the traditional tiled roof of the neighboring courtyard house to the west. The private quarters are organized around four courtyards of different sizes and orientation. This variety continues the attempt to introduce daylight from the east and the west in the rooms, so that the atmosphere of various spaces will be appreciably different at different times of the day.

Structural elements of the house are kept out of view by setting the thickness of walls to 200 mm so that columns and in-fill walls are indistinguishable once they are rendered. The maximum span of 8 m on one side of the southern courtyard required a beam that is 600 mm deep. However, this and other beams of various dimensions are kept unobtrusive by placing them above rather than below roof slabs. By subordinating structural expression, light and spatial effects, and the client's collection of furniture are allowed to draw one's attention.

这是为一对画家夫妇设计的私人住宅。基地位于一个古村的东侧边缘，毗邻大片田野。这座新建院宅的内向性与其周边传统院落的内向性保持一致。草筋白（石灰混合草筋的外墙处理，是大理地区常见且廉价的一种外立面材料）的使用也将新建筑与其周边环境联系起来。

这座房子被设计为南北两个区域，北部是夫妇二人的私密区域，南部是一组生活起居空间、客房以及保姆的卧室。9 个不同大小的院子缩小了传统院落住宅的庭院尺度，并对应着不同的房间。房子主体部分的动线将院落与一系列的功能房间联系起来。位于西南处的入口序列包含两次 180° 的转向，引领访客到达南部的前庭。西侧的一条长廊院可直接到达私密区域，途中还能一瞥中庭的水面。从南侧的中庭，可经过一个围合的用餐区，径直走向水池。储藏室模糊了起居空间的尺度感，从起居空间望出去，向东是大片的稻田，向西则是中庭的水池以及西侧廊院内的框景和邻居院落的瓦屋面。后庭为私密区域，围绕着 4 个尺度、朝向各不相同的天井来组织。因此，阳光从不同的方向被引入建筑内，人们可以察觉到空间氛围在一日之内随时间的变化。

对这座房子而言，结构元素不希望被显现。设计师将墙体厚度设为 200mm，墙体被刷上外墙涂料后人无法分辨柱子和填充墙。在南部子中，有一边的最大跨度达到 8m这就要求梁高要达到 600mm。然而不同尺寸的梁因被反置在屋面板之而变得不可见。在这里，结构的呈是次要的，光影以及业主收藏的家和物件得以成为营造氛围的主体。

pp. 102–103: View of the pond towar the south. The glazed dining area is seen on the right, and the main living area on the left. This page: Aerial vie of the house. Photo by the architect. Opposite: View of the pond toward the west. Tiled roof of the neighborin house can be seen above the wall. All photos on pp. 102–105 except as note by Chen Hao.

102–103页：朝向南看水池。可看到右边玻璃围合的用餐区和左边的主要生活区本页：鸟瞰住宅。对页：朝向西看水池从围墙的上端可看到邻居家的瓦屋面。

Credits and Data
Project title: Artist's House in Xizhou
Client: Meng Zhong, Wen Yi
Location: Xizhou, Dali, Yunnan, China
Design phase: Aug. 2014 – Feb. 2015
Construction phase: Feb. 2015 – Jan. 2016
Architect: Zhaoyang Architects
Design team: Zhao Yang, Shang Peigen
Structure system: Concrete bearing wall + concrete block
Floor area: 426 m^2
Cost: RMB 2,000,000

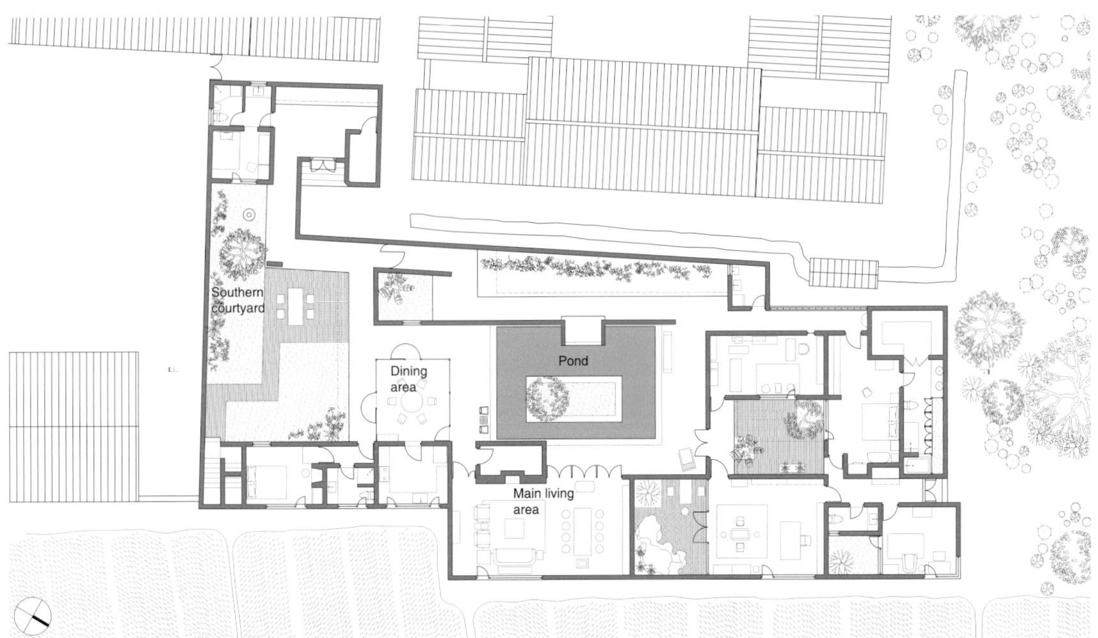

Plan (scale: 1/400) / 平面图（比例：1/400）

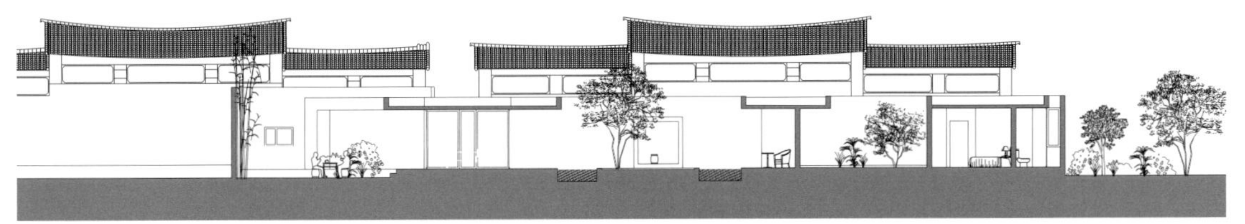

Section (scale: 1/400) / 剖面图（比例：1/400）

Atelier FCJZ

Museum Bridge

Anren, Sichuan, China 2012

非常建筑

桥馆

中国，四川省，安仁镇 2012

Feature:
Architects in China

Atelier FCJZ
Museum Bridge
Anren, Sichuan, China

The Museum Bridge is, as the name suggests, not only a museum building in Jianchuan Museum Town but also constitutes as part of the urban infrastructure – a foot bridge.
In our design, the bridge is considered as a component of urban public space and the continuity of streets. The museum belongs architecturally to the bridge while serving as a connection of urban fabric on both sides of the rivers. Therefore, strategically, the Museum Bridge is two independent structures married into one. It embodies both the characteristic anchoring tendency of a museum and the typical gravity-defying gesture of an arch bridge. Therefore, this is architecture that juxtaposes the qualities of museum and bridge and is both light and heavy at the same time. As urban public space, the arch bridge provides two pleasurable experiences – crossing the river and hanging around. The bridge is lined with public seats that allow a moment of relaxation and its surface is shaped in a curvilinear way that connects the street and the square.
In order to strengthen the theme of "Ten-Year Cultural Revolution, " the museum's formal language is designed to reflect the classical order of the Chinese architecture during the 1960s to 1970s but updated with contemporary spatial organization and structural expressions. The museum is divided into two floors – closed box on the lower floor and open pavilion on the upper floor. The exhibition is divided into six halls by glass light wells that go through both floors. The dim space on the lower floor represents the first six years of the turmoil of the "Cultural Revolution" while the bright upper floor the hopeful last four years. The two glass light wells pierce through the exhibition hall at an angle ensuring perennial sunshine downward and light on the silk screen prints of giant Mao images. Functionally, the wells double as the roof drainage. On the upper floor, there is a teahouse as well as outdoor platforms that provide visitors with the view along the river as well as the panorama of Jianchuan Museum Town.
Given the topography, the building is supported by 13 tilted concrete columns. The bridge and the first floor are made of reinforced concrete, and the second floor is made of a light steel frame system. The Museum Bridge's facade materials include exposed concrete molded with bamboo formwork, wood sandwiched walls, and wooden framed glass, taking advantage of the locally abundant bamboo and wood resources. The large wooden doors can be totally opened up. The material choice of the Museum Bridge shows the contrast and integration of two dramatically different technological systems – contemporary civil construction and local building traditions.

Credits and Data
Project title: Museum Bridge
Client: Sichuan Anren Jianchuan Culture Industry Development Ltd.
Location: Anren, Dayi, Sichuan, China
Design Period: 2009–2010
Completion: 2012
Design: Atelier FCJZ
Principal: Yung Ho Chang
Project manager: Liu Lubin
Project team: Wu Xia, Guo Qingmin, Liang Xiaoning, Feng Bo
Cooperation: Shenzhen Xinzhongjian Architecture Design Consultancy Ltd.
Site area: 2,403.2 m^2
Building area: 2,114 m^2
Building stories: 1
Building height: 5.2 m
Construction type: Reinforced concrete mega framework
Main material: Concrete with bamboo formwork

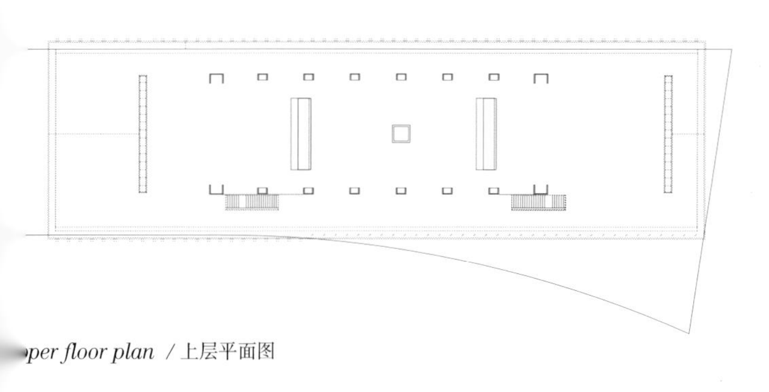

per floor plan / 上层平面图

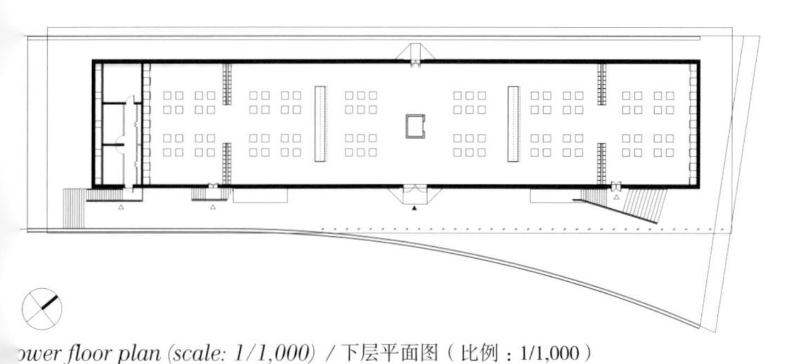

ower floor plan (scale: 1/1,000) / 下层平面图（比例：1/1,000）

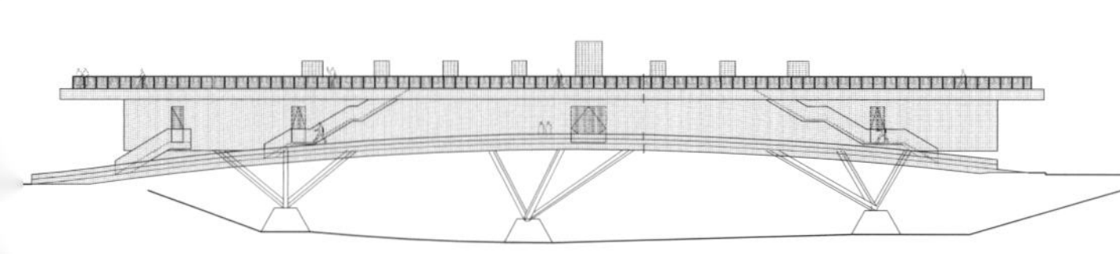

ast elevation (scale: 1/1,000) / 东立面图（比例：1/1,000）

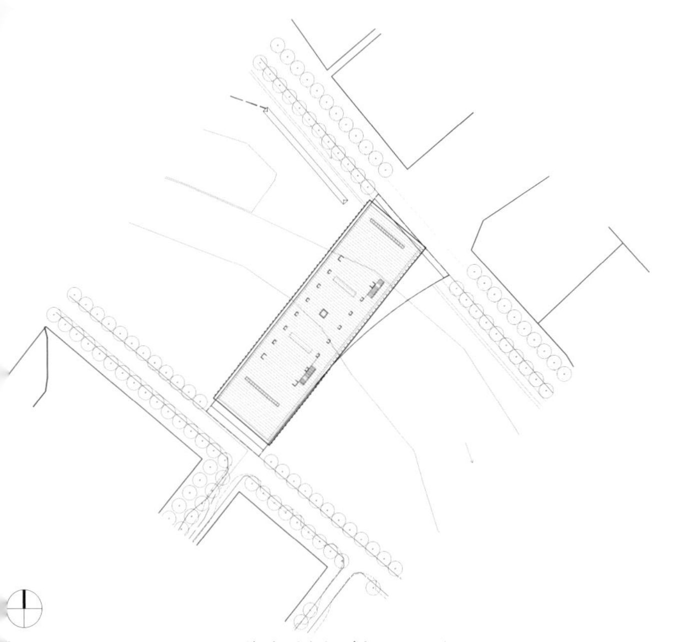

Site plan (scale: 1/2,500) / 总平面图（比例：1/2,500）

桥馆是四川安仁建川博物馆聚落规划建设的博物馆之一，同时也作为基础设施——一座桥，来使用。

桥被看作是城市公共空间的一部分，是街道的延续，而桥馆既是桥的组成元素又是两岸城市肌理的联系。因此在策略上，桥馆是两个独立结构的相加而构成的一个整体，同时具有稳重的博物馆和拱起的薄拱桥两种品质。因此，桥馆是一栋既轻又重、兼具博物馆和桥特性的建筑。

拱形桥作为城市公共空间带来过河和游逛两种乐趣。桥平面呈连接街道与广场的弧线，桥上沿河成排布置公共座椅，供行人休憩。

为了突出“十年文革”的主题，桥馆在形式语言上呼应着20世纪六七十年代中国建筑的古典格局，但对其空间组织和结构逻辑进行了重构。桥馆分上下两层，一层是封闭的“箱”，二层是开敞的“亭”。中心的交通核、两层通高的玻璃光井和展柜，将一层展览空间分为6个展厅，昏暗的下层空间对应黑暗的“文革”前6年，明亮的上层空间代表充满希望的后4年。两道通高的玻璃光井斜穿展馆，其斜度刚好保证一年四季都有阳光顺井而下，照亮丝网印制的毛主席巨幅画像。在功能上，光井还兼为二层屋面的排水井。二层还设有茶社、室外平台，为参观者提供休闲、观赏河道景观和建川博物馆聚落的视野。

为适应地形，在建筑底层设置了13组不同斜度的混凝土束柱。桥和一层建筑是钢筋混凝土结构，二层是轻钢剪力墙混合结构。桥馆立面使用了小竹模清水混凝土、双层夹芯木板墙、木肋玻璃墙，建造上利用当地丰富的竹、木资源。展馆拥有可完全开敞的大面积木板门。桥馆建筑材料的选择体现了当代市政建设与传统地方建筑两大不同技术系统的对比和统一。

pp. 106–107: View of the staircase leading up to upper floor. Opposite: View under the bridge. This page, above: General view from the northeast. This page, below: Distant view from the riverside. All photos on pp. 106–109 by Cao Yang, courtesy of the architect.

106–107页：通向上层的楼梯。对页：桥下的景象。本页，上：从东北侧看到的全景。本页，下：从河畔远看桥馆。

Wang Hao

Wang House

Ningbo, Zhejiang, China 2012

王灏

王宅

中国，浙江省，宁波市 2012

Site plan (scale: 1/1,200) / 总平面图（比例：1/1,200）

Wang Hao
Wang House
Ningbo, Zhejiang, China

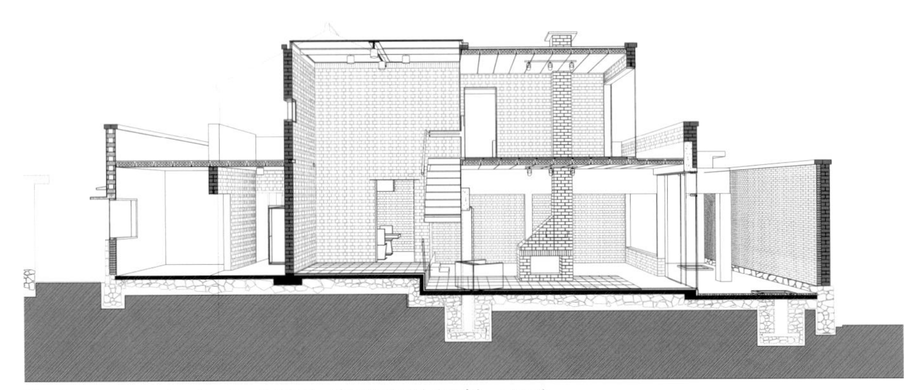

Section perspective (scale: 1/200) / 剖面透视图（比例：1/200）

pp. 110–111: General view of the west facade. Bricks from the old houses are reused in certain parts such as the perimeter wall. pp. 112–113: View toward saloon. All photos on pp. 108–113 courtesy of the architect, except as noted by Liu Xiaoguang.

110–111页：西侧外立面的的全景。老房子的砖被重新利用在某些部分，比如外墙。112–113页：看向客厅。

The residence is located in a typical coastal village at Chunxiao Town, Ningbo City, where the architect spent his childhood. This residence is built based on extension of two existing farmhouses.

This residence adopts red bricks and precast slabs made locally. It serves as a place for holidays and temporary studio. Its overall layout utilizes a centripetal plane, and heavy brick walls imply a strongly internalized residence philosophy. All building materials and furniture are made of common materials in the country to express the simplest view of life. It adopts brick-concrete structure and its framework is completely free. In detail, beam columns dissociate between walls and floors (precast slabs), which builds flowing structured space. The contrast between linear composition shown by internal space and walls in the external space happens to coincide with the internal and external duality aesthetics of ancient dwellings in China.

Three clip-shape cross walls and the centripetal layout guide space to go towards the inside hierarchically. An internal courtyard in the centre, two floors in height, acts as a core to organize daily functions to trace back to traditional "internalized" space experience. Three cross walls, which are arranged in the form of ascending order according to their height, build a kind of ancient closed architectural appearance in the direction of the west wall, and also show the horizontal tension of this residence by their simple and distinct outline. As response to the the situation that a large area appears closed, the structure of the newly built part insidc thc rcsidcncc adopts a form of free organization. Specifically, slabs, beams and columns are handled separated as independent elements to intersperse them among walls. Traditional lap joint organization methods are used to transfer gravity freely. Flexible columns adjust and divide functional space. At the same time, different sections and beams with different heights enhance the spatial level and sense of segmentation and build structured flowing space jointly. Traditional woven bricks exist in the form of filling and original walls and sections are inserted continuously (old residence before transformation). As a result, when the purity of the walls is weakened, the dominant position of the free framework as a subject is strengthened as well. A sense of beauty and intuition about simplicity of original materials are kept and shown completely. As time goes by, the pigment of bare red bricks accumulates gradually and forms a stable visual balance together with the plain concrete and water-milled ground. Some traditional rural old furniture, such as hand-made bamboo chair, brick sofa that imitates the brick "fire pit" in the countryside, straw rain cape and kitchen after improvement in brick cooking bench, build the atmosphere of farmhouses. Binary opposition between the large external wall space and internal free linear structure comes from the most ordinary knowledge about aesthetics of ancient dwellings in China, i.e., internalization.

该住宅位于宁波市春晓镇，一个普通的滨海小村，是建筑师王灏童年生活的地方。它是在原有两间农舍的基础上扩建而来的。

该住宅作为度假和临时工作室使用，材料上使用了红砖和当地普遍的预制混凝土板。总体布局采用了向心性平面，厚砖墙强调了居住空间的封闭性。所有的建筑材料和家具均采用当地常见的材料，以传达一种平和朴素的生活观。建筑结构为砖混结构，自由框架。梁柱与墙体和楼板分离的设计形式，创造出一种流动的结构空间体验。内部空间的线性与外部空间的墙体封闭性相映成趣，并与中国古代内、外二元性的住宅美学相契合。

3道回形横墙，向心式布局，引导空间向内层层递进，以中央两层通高的内天井作为核心组织日常功能，回溯到传统的"内向"空间经验。3道从低到高的横墙主要在西山墙方向营造一种自古以来的封闭建筑外观，亦以极简的明确轮廓勾勒出该住宅水平向的张力。作为对外观大面积封闭的回应，内部新建部分的结构采用自由组织方式，把楼板、梁和柱子作独立元素分离操作，穿插在墙体之采用传统搭接的组织方式，自由传重力。灵活的柱子调节和分割了各功能性空间，各个不同截面不同高的梁加强了空间的层次感和分割共同营造出结构化的流动空间。传的织体型材料砖以填充的形式存并有原来的墙体和部分结构（改造的老宅）不断地插入，局部拆下来老砖亦被随机加入进来，在墙体被弱纯洁性的同时，加强了自由框架为主体的支配地位。

原始材料的美感和朴素直觉保留下来，并全部得以呈现。裸露红砖随时间的消逝，色素沉淀，与色混凝土和水磨汀地面形成稳定的觉平衡。一些老的传统农村家具，手工竹椅、模仿乡下砖砌"火坑"砖砌沙发、蓑衣和砖灶台改良后的房，营造了农舍的氛围。外部大墙与内部自由线性结构的二元对立来中国古民居最普通的美学常识——部化。

Credits and Data
Project title: Wang House
Location: Ningbo, Zhejiang Province, China
Design period: 2009–2010
Construction period: 2010–2012
Construction engineer: Hong Wenming
Construction team: Villagers and local craftsmen
Silk scroll: Shi Jie
Site area: Approx. 260 m²
Total floor area: Approx. 220 m²

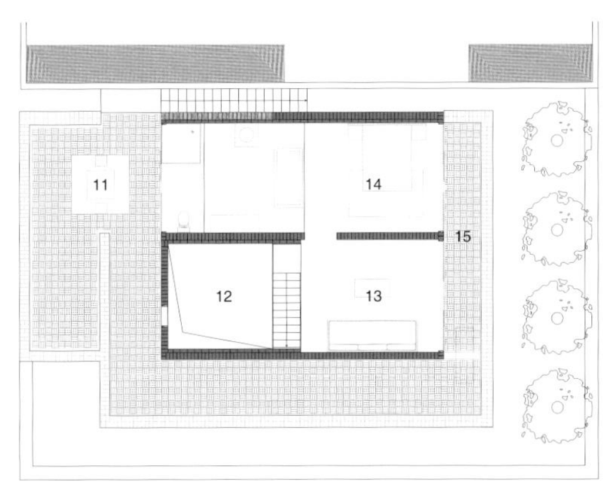

Second floor plan / 二层平面图

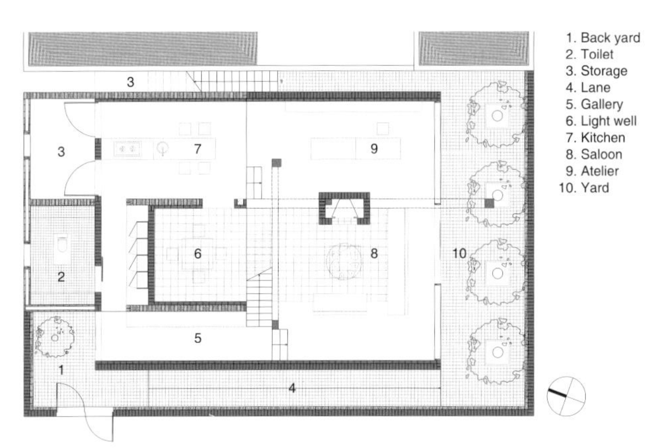

Ground floor plan (scale: 1/300) / 首层平面图（比例：1/300）

Opposite, 2 images: Concept images. This page, above: View of the bedroom on the second floor. Photo by Johan Sellén. This page, below: View of the light well on the ground floor.

对页，两张：概念图。本页，上：二楼的卧室。本页，下：看首层的天井。

Li Xiaodong Atelier

LiYuan Library

Huairou District, Beijing, China 2011

李晓东工作室

篱苑书屋

中国，北京市，怀柔区 2011

Feature:
Architects in China

Li Xiaodong Atelier
LiYuan Library
Huairou District, Beijing, China

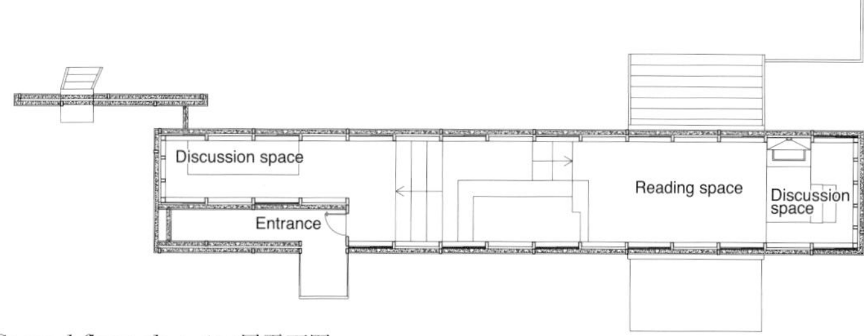

Second floor plan / 二层平面图

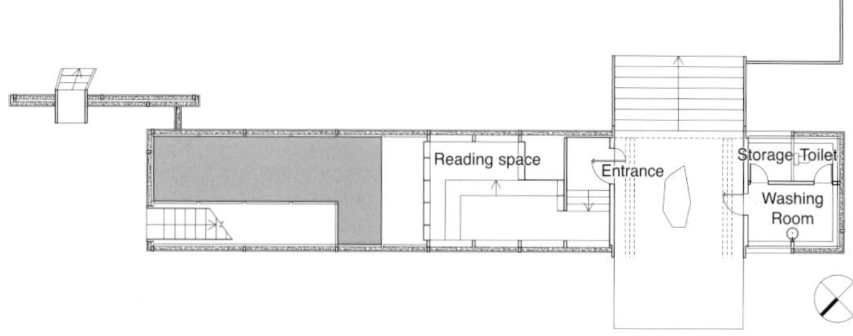

Ground floor plan (scale: 1/400) / 首层平面图（比例：1/400）

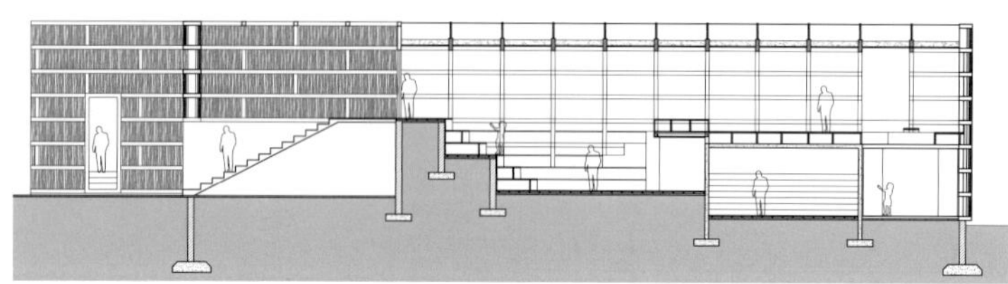
Section (scale: 1/400) / 剖面图（比例：1/400）

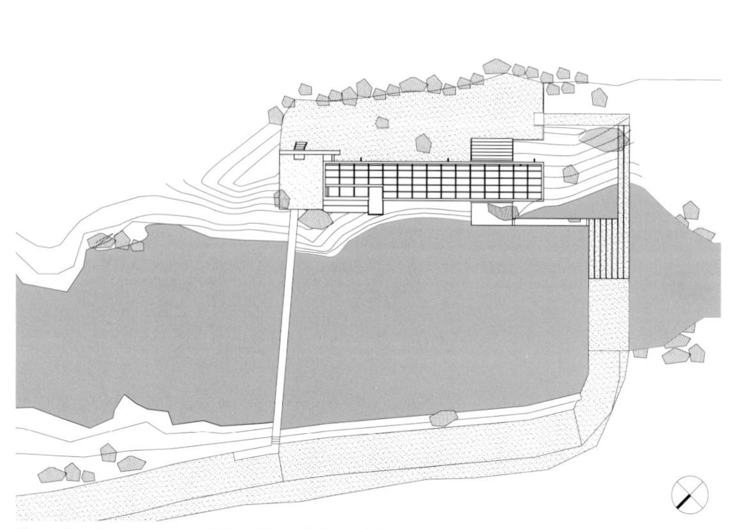
Site plan (scale: 1/1,000) / 总平面图（比例：1/1,000）

This project is a modest addition to the small village of Huairo on the outskirts of Beijing, just under a two-hour drive from busy Beijing urban life.
On the one hand it forms a modern programmatic complemen to the village by adding a small library and reading space with a setting of quiet contemplation. On the other hand we wanted to use architecture to enhance the appreciation of the natural landscaping qualities.
So instead of adding a new building inside the village centre, w chose this particular site in the nearby mountains, a pleasant five minute walk from the village centre. In doing so we could provide a setting of clear thoughts when one consciously takes the effort to head for the reading room.
Because of the overwhelming beauty of the surrounding nature our intervention is modest in its outward expression. We can't compete with nature's splendor. The building blends into the landscape through the delicate choice of materials and the careful placement of the building volume. Especially the choice of material is crucial in blending with the regional characteristics. After analyzing the local material characteristic in the village we found large amounts of locally sourced woode sticks piled around each house. The villagers gather these stick all year round to fuel their cooking stoves. Thus we decided to use this ordinary material in an extraordinary way, cladding th building in familiar textures in a way that is strikingly sensitive The inside of the building has a very expressive character though; its interior is spatially diverse by using steps and small level changes to create distinct places. It frames views towards the surrounding landscape and acts as an embracing shelter. The building is fully glazed to allow for a fully daylight space. The wooden sticks temper the bright light and spread it evenly throughout the space to provide for a perfect reading ambience.

篱苑书屋是北京市怀柔区一个小村庄的增建项目，从市内驱车两小时即可到达这个位于京郊的谦和空间。

在清谧的自然景观中添置一个小书屋以及可供人们静心阅读的空间，既补足了村庄的现代气质，又通过“消隐”的建筑介入手法，强调了自然环境的价值。

设计师将建筑地段选择在村庄附近一处脚力可及且依山傍水的山中荒地，而非在村子中心强行兴建，这样就表达了我们有意识地将人们引入书屋的明确动机。

周边的如画风景难出其右，因而建筑的介入姿态谦逊。建筑通过对材料和地段的精心选择，微妙地融入自然景观。尤其是材料的选取具有很强的当地特色。村庄中房前屋后总会堆些收集来的柴火棍，这是村民们用来做饭的燃料。书屋正是将这种习以为常的材料以特别的方式来加以利用的。用这样的材质包裹建筑，却得到了令人震撼的效果。

在谦逊的外形之下，建筑的内部空间有很强的表现力。通过台阶的运用和高度的变化形成分区，营造出丰富的空间，而完全由玻璃封闭的框架使内部在整个白天都能自然采光。柴火棍将明亮的光线折射分散，使空间中的光线均匀且柔和，从而创造出理想的阅读环境。

pp. 116–117: View of the entrance on the northeast. Opposite: Interior view. All photos on pp. 116–119 courtesy of the architect.

116–117 页：东北侧的入口。
对页：建筑室内。

Credits and Data

Project title: LiYuan Library
Client: Jiaojiehe Village
Commissioning donors: Luke Him Sau Charitable Trust
Location: Jiaojiehe Village, Huairou County, Beijing, China
Construction period: March 2011– October 2011
Architect: Li Xiaodong Atelier
Team: Li Xiaodong, Liu Yayun, Huang Chenwen, Pan Xi
Floor area: 175 m^2
Construction cost: RMB 1,050,000

URBANUS

Urban Tulou

Foshan, Guangdong, China 2008

都市实践

土楼公舍

中国，广东省，佛山市　2008

Credits and Data

Project title: Urban Tulou
Client: Shenzhen Vanke Real Estate Co., Ltd.
Location: Foshan, Guangdong Province, China
Design period: 2005–2007
Completion: 2008
Architect: URBANUS
Design directors: Liu Xiaodu, Meng Yan
Technical director: Zhu Jialin
Project directors: Li Da, Yin Yujun
Designers: Huang Zhiyi, Li Hui, Cheng Yun, Huang Xu, Zuo Lei, Ding Yu, Wei Zhijiao, Li Jing, Wang Yajuan, Zheng Yan, Shen Yandan
Interior design: Guoqun Studio
LDI: Archilier Architecture LLC
Logo design: Huangyang Design
Site area: 9,141 m^2
Total floor area: 13,711 m^2

建筑与都市
Architecture and Urbanism
Chinese Edition 16:06

064

Feature:
Architects in China

URBANUS
Urban Tulou
Foshan, Guangdong, China

pp. 120–121: Aerial view from the northwest. Photo by Yang Chaoying, courtesy of the architect. This page: View of the entrance with shops and restaurants on the ground floor. Photo by Iwan Baan. Opposite: Hekeng Tulou cluster, one of the Tulou groups in Fujian Province. Photo courtesy of the architect

120–121 页：从西北侧空中俯瞰建筑。
本页：一层入口的商铺及餐厅。
对页：福建省的河坑土楼群。

lou is a dwelling type unique to the the Hakka people. It is :ommunal residence between the city and the countryside, tegrating living, storage, shopping, religion, and public tertainment into one single building entity.
aditional units in Tulou are evenly laid out along its rimeter, like modern slab-style dormitory buildings, but with eater opportunities for social interaction. By introducing a ew Tulou" to modern cities and by carefully experimenting ; form and economy, one can transcend the conventional odular dwelling into urban design. Our experiments explored ays to stitch the Tulou within the existing urban fabric, which cludes green areas, overpasses, expressways, and residual land ft over by urbanization. The cost of residual sites is low due incentives provided by the government; this is an important ctor for the develoment of affordable housing. The close oximity of each Tulou building helps insulate the users from e chaos and noise of the outside environment, while creating ı intimate and comfortable environment inside.
tegrating the living culture of traditional Hakka Tulou uildings with affordable housing is not only an academic sue, but also implies a more important yet realistic social ıenomenon. The living conditions of impoverished people is ow gaining more public attention.
he research of Tulou dwelling is characterized by omprehensive analyses ranging from theoretical hypothesis to ractical experimentation. The study examined the size, space atterns, and functions of Tulou. The new programs also inject ew urban elements to the traditional style, while balancing the ension between these two paradigms. As a consequence of such omprehensive research, the Tulou project has accumulated ayers of experiences in various aspects. The project provided platform for an in-depth discussion on feasibilities and ossibilites of contextualizing the variable metamorphoses f traditional dwelling modules with an urban reality. It also ntroduced a series of publications and forums on future ypothetical designs for a "new Tulou project". The logic and esign process of the Tulou program set up a solid foundation nd excellent precedent for translating research-based feasibility tudies to design realization.

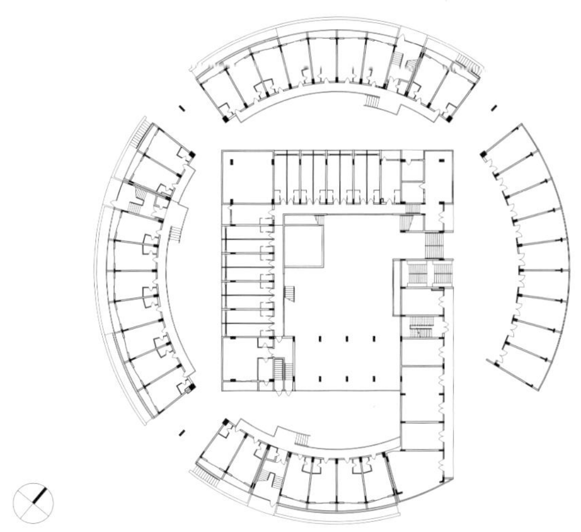

Ground floor plan (scale: 1/1,200) / 首层平面图（比例：1/1,200）

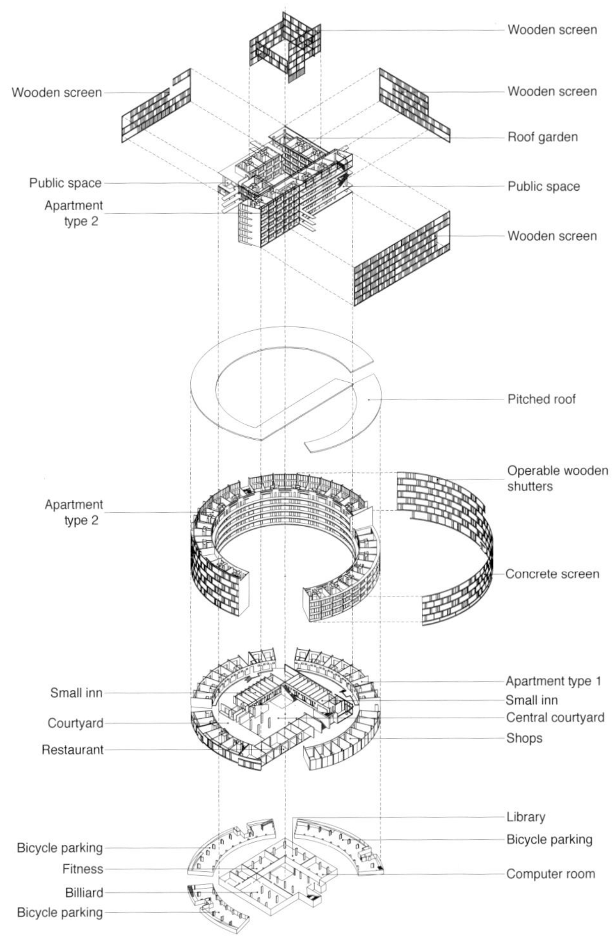

Exploded axonometric drawing / 展开的轴测图

土楼是客家民居独有的建筑形式。它用集合住宅的方式，将居住、亡藏、商店、祭祀和公共娱乐等功能集于一个建筑体量中，具有巨大凝聚力。

本例将土楼作为当前解决低收入人群居住问题的方法，而不只是形式上的承袭。它与现代宿舍建筑类似，但又具有现代走廊式宿舍所缺少的亲和力，有助于保持低收入人群在社区中的邻里感。将“新土楼”植入当代城市的典型地段，与城市空地、绿地、立交桥、高速公路、社区等典型地段拼贴。这些试验都是在探讨如何利用土楼这种建筑类型去消化城市高速发展过程中遗留下来的不便使用的闲置土地。由于获得这些土地的成本极低，从而使这类经济适用房的开发成为可能。土楼外部的封闭性可将周边恶劣的环境予以屏蔽，同时内部的向心性又创造出温馨的小社会。

将传统客家土楼的居住文化与经济适用房结合在一起，标志着低收入人群的居住状况开始进入大众视野。

这项研究的特点是分析角度的全面性和从理论到实践的延续性。对土楼原型进行尺度、空间模式、功能等方面的演绎，然后加入经济、自然等多种城市环境要素。在多种要素的碰撞之中寻找各种可能的平衡，这种全面的演绎保证了丰富经验的获得，并可为深入的思考提供基础。从调查土楼的现状开始，研究传统土楼在现代生活方式下的适应性，将其“城市性”发掘出来，然后具体深化，进行虚拟设计，论证项目的可行性，最终将研究成果予以推广，这样从理论到实践的连续性研究，是“新土楼”构想的现实性和可操作性的完美结合。

Discussion:
Generational Development of Architecture

Philip F. Yuan, Yu Ting, Liu Yichun, Shui Yanfei
Moderators: Li Xiangning, Ma Weidong

对谈：
建筑的时代性发展
袁烽 俞挺 柳亦春 水雁飞
主持者：李翔宁 马卫东

—Shanghai is one of the cities in China where the architecture industry has developed over time, and has uniquely developed with its exclusive historical and geographical background. In this interview, four architects, Liu Yichun, Philip F. Yuan, Yu Ting, and Shui Yanfei, were invited to discuss their opinions on the Chinese architecture industry, which has been developing over time, as well as realistic problems commonly encountered in China. Philip F. Yuan is the director of an architectural office and is also professor at Tongji University. Liu Yichun and Yu Ting have worked for the Design Institute. Shui Yanfei studied abroad and worked at an architectural office in the U.S. The four architects with different backgrounds have different responses to changes and understanding of architecture. The interview consists of three parts: "Architectural Policy under a Changing Environment", "Shanghainese Architects", and "Primitive Thoughts in Architecture".

Architectural Policy Under a Changing Environment

Philip F. Yuan (PFY): During our growth process, we continue self-regulation. With structural changes in the economy lately, the roles of architects are changing as well. Our policy for adapting to the current environmental changes is "we can respond to any changes by not changing". We are hoping to find our place on the world stage, through the tireless efforts of the architecture industry.

Yu Ting (YT): Current society can be described as selective. When the industry is facing winter-like hardships, emphasis on quality must be further enhanced.

Liu Yichun (LY): The economy is sluggish today, but it could be an opportunity for the Chinese architecture industry to create better works. Many architects design with a quite basic method of using limited resources, and the buildings made in such a way encourage us to think about the basics of architecture.

YT: An architect once said, "Practical architects are short in China. Most of the existing architects are various types of commercial architects", but such fads became a thing of the past. Today, a certain number of architects exist and the one-time "pseudo-architects" are gradually growing into "segmentalized architects." I expect that diversified scenes will emerge in the future architecture industry, where each architect has his own direction. In the next issue of Chinese architecture, the current "kaleidoscope" kind of works will perhaps disappear, and characteristic architects, converted from the "pseudo-architects", will more or less increase.

Shui Yanfei (SY): There are two key points for us. The first poi[nt] is opportunities in construction projects. All offices (that were opened) in the 1980s were in a start-up phase, and one of our important project sources at an early phase is often "outside-the system" buildings, which are derived from China's unique land system. Such examples include: "renovation and rebuilding", which can be regarded as using up stocks in cities; and "farming village construction", lead by a boom in resort tourism lately. The second key point is more equal distribution of information. Information disclosure through the Internet was indeed experienced when our generation was in school. In addition, recognition of architecture by the general public is also much improved.

LY: Over the past few years, many buildings in China were changed to some extent after completion. A major cause is fewer chances for Chinese architects to have proper clients and buildin[g] occupants. This issue is becoming obvious lately. For example, there are cases of government buildings in which the future use of a building is undecided, or the details such as specifications and functions are unknown even when the building has a specifi[c] use. In these cases, if there is any opportunity, architects can participate from the early phases of the project.

YT: Based on several latest projects, I feel like the entire process i[s] delegated to me. In such cases, the clients have no clear opinions at the beginning of the project, requiring support for organizing commercialization possibilities of the building and business types In this situation, architects have to learn things covering a broad range of topics.

PFY: We also have a little different policy of "making things simple as much as possible". I believe that partnerships with proper partners are important, and I am trying to enable manifestation of everyone's abilities in the area of their specialty, while valuing excellent frameworks (reliable management companies and promoters) when proceeding with a project. There are two characteristics in our representative works in recent years. One is that very good works were completed when we could confront our inner selves. Another is that good works were completed when we could confront the creating process. When looking at the future, I think it is important to change many people's understanding of architecture by looking for new technologies or methods, rather than changing ourselves.

SY: Today's construction plans in China are hard to be connected with proper designs. In fact, the concept of consumption has led to wrong directions due to various tricks. New breakthroughs in

—上海是中国建筑发展与时俱进的城市之一，在特有的历史地理背中已形成了独有的脉络。本次上海对谈我们邀请到柳亦春、袁烽、挺、水雁飞四位建筑师，就因时代的变化而不断发展的中国建筑以在中国经常遇到的一些实际问题谈一谈各自的看法。这四位建筑师有各自不同的建筑背景，袁烽在主持事务所的同时，也是同济大学教授；柳亦春和俞挺都曾有大设计院的工作经历；水雁飞曾留学美并有在美国事务所工作的经历。他们应对变化和对建筑的理解也各不同。对谈分为三个部分：环境变化中的建筑策略、上海建筑师、筑本原思考。

环境变化中的建筑策略

袁烽：我们的成长是一个不断自我调整的过程。随着现今经济的结构转型，建筑师所扮演的角色也在变化，应对现在的环境变化，我们的策略是"以不变应万变"，希望在建筑本体上下功夫，从而在国际上找到自己的位置。

俞挺：现在的社会是一个淘汰的社会，当寒冬来临，我们应该更倾向于"质量为胜"。

柳亦春：现在经济下行了，但这也许是中国建筑的一个好时候，可以抓住机遇创造一些好的作品。由于资源的限制，很多独立建筑师都在用非常基本的方法做设计，而这种使用基本的方法来营造建筑的效果，恰好是在思考着建筑最本原的问题。

俞挺：曾有建筑师说过"中国缺乏实践建筑师，现在基本都是各种类型的商业建筑师"，但当这段浪潮过后，建筑师的数量积累到一定程度，曾经的"角色建筑师"就会逐渐向"细分建筑师"发展，未来的建筑界就会出现多样化的景象——大家各自找到自己的方向。也许做下一本中国建筑专辑的时候，就不会是现在作品呈现"万花筒"的状态，而是有多少"角色建筑师"转向了特色建筑师。

水雁飞：对我们来说有两条线索。一是项目机遇。生于80年代的建筑师所主持的事务所现在基本都在起步阶段，中国特有的土地制度导致的"体制外"的建造，往往成为我们初期重要的项目来源之一。比如城市中存量消化的"更新改造"，或者最近比较热门的休闲旅游带来的"乡建"。二是信息的扁平化。随着互联网的普及，我们这一代在读书的时候对"资讯的开放"，体会是比较深的，另一方面它也积极地影响了国内大众对于建筑的认知。

柳亦春：前些年，国内的很多建筑建成后都存在一定的变动，其中一个原因是，中国的建筑师很少能碰到真正的业主，也就是使用者。最近这些问题更为凸显，比如政府想盖房子，但不一定知道要盖的房子未来具体的用途，或者知道用途但不知道具体的功能配置，此时如果建筑师有机会介入，就会参与到项目的前期。

俞挺：从我最近做的几个项目来看，我觉得我的角色好像变成了全过程的holding。所谓"全过程holding"，就是业主在刚开始时没有任何明确的角度，需要你帮他理清可行性、业态等问题。在这种情况下，建筑师要扩展学习的东西是很多的。

袁烽：我们的策略略有不同——尽可能把事情做得简单。我认为找到合适的人合作很重要，在工作中我比较倾向有一个很好的架构，包括有很好的设计管理公司、甲方等，我们都在做自己擅长的事情。这几年来，我们的代表建筑作品主要有两个特点，一是真正面对自我内心的时候，可以做好；二是真正面对工艺的时候，可以做好。所以，在未来我们可以把控的也是这两方面，就是我们能否寻找一种通过新工艺的方式来改变更多的人对建筑的理解，而不是改变我们自己。

水雁飞：目前国内的策划难以对接真实的设计，相反却往往会设下圈套，误导市场的消费观念。建筑师合理的前期介入，提供有洞察力的观点，让运营和使用模式与设计关联起来，也可为设计在类型上的突破提供一些前提。

上海建筑师

俞挺：其实"上海人"不是一个地域概念，而应是一个文化概念，由特殊的地理历史造就的特殊文化，只要融入这个文化，就是上海人。所以"上海建筑师"更多的是一种建筑文化的界定吧。

柳亦春：上海有非常好的街道感和空间尺度，也许是中国最好的城市空间遗产。这些都会很好地、潜移默化地熏陶一个建筑师的成长。大舍的作品基本是立足扎根于上海这块土地的，那么就必须深入上海的

design will be found by the participation of architects at the early phases of a project, connecting designs with building operation and intended uses, and by proposing ideas with insight.

Shanghainese Architects

YT: So-called "Shanghainese people" are not based on a geological attribute, but a cultural attribute. There is a geographically and historically unique culture in Shanghai. People who assimilate with the culture are Shanghainese people, and therefore "Shanghainese architects" can be defined as a certain type of architectural culture.

LY: Shanghai has street corners and spaces with a very nice atmosphere, and one could say they are the best urban-space ruins in China. These elements make an ideal place for architects to be trained. Most works by Atelier Deshaus are rooted in the land of Shanghai, deeply penetrate into the Shanghai culture, absorb nutrients from the city, and become buildings that fit into the area. Over the course of this process, there is a necessity to recognize the problem that many of the excellent spatial heritages were built during the colonial period, and how architects should establish a new standpoint through their work.

PFY: A solo exhibition by Ding Yi at the Long Museum was particularly impressive. He has been tackling one theme for decades, though there were surely directional corrections along the way. I feel that his works have considerable persuasiveness and strong influence.

LY: I agree. He has been continuously working on the same theme and is absolutely committed to his works, while never being influenced by others. Today's consumption society often makes us lose sight of the true nature of things.

PFY: In fact, architects – whether in Beijing or in Shanghai – are facing various problems in their growth process, but interestingly the problems are different from each other. Consequently, rather than comparing architects based on good/bad or strong/weak, we should look at differences in growth of each architect over a longer time period – 10, 20, or 30 years – to find small changes.

YT: A historian said, "one should continue to wait for 100 years to solve all historical problems, until all stakeholders are gone". Therefore, valuable information might be found when looking for hints out of necessity after 100 years, at a time when the problems are forgotten. We might be rushing too much for a conclusion.

LY: I think my career has two values. One is, as an architect, creation of spaces that produce a certain level of happiness and comfort. Another is establishment of criticism against the time and the society (i.e., retaining clear thoughts under a given environment), and I recognize the latter is more valuable. However, the basis is provision of services as an architect, based on my skills and expertise.

Primitive Thoughts in Architecture

YT: In my opinion, knowledge < plausible truth < intelligence < chatter. In this age, we can understand the importance of knowledge, intelligence, and plausible truth. Although we can gain an insight on the movements of society through chatter, this is different from criticism. I sometimes hide the standpoint of an "architect", because the world cannot be viewed well from architect's standpoint at times.

PFY: Currently, China is in a state of excessive consumption and information exchanges. For example, in the magazine Log 30, Pia Ednie-Brown described the modern age as "nothing happening", and that "voice and movement are large, while ste for advancement are fairly small. Opinions are exchanged and new topics are created by maximum effort, but actually, nothin is happening". There are things accompanied by significant suffering, but we can truly advance because of such things.

SY: In the U.S., making good use of critical thinking in the field of art ultimately reconstructed the voice of art. This derived a voice for the architecture field, in addition to in art-related areas. However, what makes architecture different from art is the presence of people's recognition of spaces; in other words, the permanent presence of "Topography, Orientation, and Physiognomy", proposed by Dalibor Vesely.

PFY: I agree with that opinion. In the architecture field, most things cannot be changed easily. Innovations in technology/art and existing cultures/processes coexist for quite a long period of time. I have been studying morphology manipulation for a few years. I believe that the most fundamental issue in architecture is research on materials, processes, and spaces. For architectural works as actually created spaces, a new essential momentum can be seen in various areas from materials, to processes, to structures, to performance.

LY. Progress from manual drawings to PCs, and to even newer technologies, will bring many new possibilities.

PFY: "Production" and "research" are closely related, but my current focus is on research.

SY: Excellent architects often avoid a certain kind of simple features (such as spaces, materials, and structures). From observation of their series of works over time, we can see how they balanced many elements, and selected openness or closeness. In the end, the diversity in their works is retained whil maintaining a constant direction in terms of attitude towards design.

YT: I listed architecture-related elements in a table by macro-thinking process. When I found the length of lifecycles of the elements, I discovered the cycle is circulating. I call this a "micrometer caliper of architecture".

LY: If a building still remains powerful 100 years later, it can be said to be a truly good architectural work. It might be described as a building with a classic dignity. However, there are good and bad in classics, and good or bad depends on the quality of the building itself, rather than its dignity.

PFY: Clues for architecture development are not necessarily found in timeless masterpieces. I think they might be found in works other than such masterpieces.

LY: With the most basic architectural method, we can create influential architectural works. The basics are always scale, rhythm, materials, structure, etc. These construct relationships between a building and its environment and between a building and the people that touch people's hearts and create comfortable architectural works.

Translated from Chinese by Haruki Makio

化，从城市中获得养分，做适合这个地域的建筑。在这个过程中，面临着其他问题——上海曾是一个殖民城市，很多优秀的空间遗产是殖民时期留下的，当代建筑师如何以作品建立自己新的身份？

烽：令我印象深刻的是丁乙在龙美术馆西岸馆做的个展，他几十年持做一件事情，虽然过程中有不同的修正，但作品很有说服力，具非常强的感染力。

亦春：对，就是不受任何影响地坚持做同一件事情，全身心地投入中。现在是消费的世界，人们的消费取决于很多方面，但事物本身有其本质，这个本质在很多时候会被无意掩盖。

烽：其实，无论是北京建筑师还是上海建筑师，每个人在成长过程，都面临着不同的问题，而最有趣的部分是各个问题的差异性。所不要用好坏、强弱去比较建筑师，应该放在历史的背景中来看问，10年，20年，30年后再来看问题，总会窥探到其成长中的一些蛛马迹或潜移默化的变化。

挺：有一位史学家说过“所有的历史问题都需留待百年之后，当所相关利益者都没有关系的时候”，所以当事情被遗忘了100年之后，今急迫想要追寻的线索也许会被发掘出更有价值的信息。现在的我太急于总结当代史。

亦春：以我的经历来看，我有两种价值，一是作为建筑师，你能带别人多少愉悦；二是你对时代与社会能建立什么样的批判性（在一环境中你如何保持一个清醒的头脑），我认为这个批判性会更加具价值。但最根本的仍然是，作为建筑师，你能够靠你的手艺和专业识服务别人。

建筑本原思考

挺：knowledge<plausible truth<intelligence<chatter，这是我的观点。现在的时代，我们可以理解knowledge，intelligence和plausible truth，而chatter却可以洞悉社会的震动在哪里，这不一定是批判性。有时我试图将“我是建筑师”的身份抹掉，因为个体差异，“建筑师”的身份有时会遮挡我看世界的视角。

袁烽：现在全中国的状态是过度消费概念、信息过度交流。即如Pia Ednie-Brown在LOG第30期上评价如今的时代，“其实什么都没有发生”，“声音和动静很大，前进的脚步却很小，花很大的力气讨论、空谈和制造新话题，但实际没什么发展”。有的事情在过度折腾，而真正能前进的还是在本体层面。

水雁飞：美国利用艺术上的批判性思维，最后可以重新建立起艺术的话语权，它自然地也衍生到和艺术相关的领域，包括建筑。但区别于艺术，建筑有一些比较恒定的东西，即人对空间的认知，就如 Dalibor Vesely 所提及的 Topography, Orientation, Physiognomy.

袁烽：我同意水雁飞的观点，建筑有很多东西不会轻易改变，通过技术、艺术等方式的创新与既有的文化与工艺在相当长的一段实践中是并存的。这几年，我做得较多的是形式操作研究。我相信，材料工艺与空间形式的研究还是建筑学重要的本体问题。建筑作为物质化的空间，在材料、工艺、结构以及性能等层面，存在着原发与内在的全新动力。

柳亦春：从手画图，到电脑，再到各种新技术，一定会带来很多新的可能性。

袁烽：“做”和“研究”是有紧密联系的，现在最关注的还是研究的部分。

水雁飞：优秀的建筑师往往会避免某种单一向度的特征，比如空间、材料或者结构等。观察他们的一系列有时间跨度的作品，可以分析他们如何“平衡”多元的基因，如何选择开启或关闭。最后保持作品的多样性，而设计态度又相对延续。

俞挺：我用一种宏观的方式，将与建筑学有关的因素列入了一个表格中，发现这些因素生命周期长短，其实是有迹可循的，我把它称为建筑学的游标卡尺。

柳亦春：一个房子过了百年之后，如果仍然会让你感到充满力量，那就是真正的好建筑。它也许是一个被称之为古典风格的建筑，但同样是古典，可以是很好的，也可以是很差的，好坏不在于风格的差异，而取决于房子真正的好坏。

袁烽：对我来说，建筑学的发展线索不见得存在于永恒的经典中，很可能存在于永恒之外的不同语境线索中。

柳亦春：我们可以用建筑最基本的方式做出有感染力的房子。最基本的永远是尺度、节奏、材料、建造等，它在于处理建筑与具体环境之间的关系、建筑与人之间的关系，做出让人心生感动和愉悦的建筑。

Zhu Jingxiang

New Bud Building System

Sichuan / Jiangsu, China 2010–2014

朱竞翔
新芽建筑系统
中国，四川省 / 江苏省 2010–2014

Since 2008 Professor ZHU Jingxiang from the Chinese University of Hong Kong (CUHK) and his team have invented several lightweight building systems with impressive architectural qualities and outstanding structural performance. New Bud Building System (NBS) is one of those and has a composite structure consisted of a light gauge steel frame strengthened by rigid wooden panels. Such a composition results in features such as light-weight and outstanding resistance to seismic forces. The multi-layered, continual envelope provides a comprehensive thermal solution, resulting in good indoor comfort. Erection takes two to four weeks with the dry method. A building constructed with this system can also be demounted, relocated or renovated easily, with very low impact to the surroundings. It is suitable for buildings of four stories or less. The surface can be customized.

After the China Sichuan 512 earthquake, the design team from CUHK applied NBS in provinces of China that are especially prone to earthquakes. The challenges posed by construction in remote locations lacking in resources pushed designers to develop flexible strategies and efficient tools, linking the rural and industrial reality of mainland China with academic research. From such confrontations has emerged a range of contemporary and localized solutions that address the general challenges of sustainable development.

All prefabricated components in the projects except Grameen Bank of the Yunus Centre were ordered from factories in Chengdu and Chongzhou, to minimize energy use in transportation and utilize local resources. This strategy guaranteed building quality under stringent budget conditions. In series projects, the team was not only responsible for the design, but also contributed to many other tasks, from budget planning to post-occupancy evaluation. Through these projects, the system has proven its wide applicability, while the team has outlined a unique approach for sustainable development, for both the user and local building industry. The built works consequently reinstate the architect's original role as both master of construction and guardian of society.

"Bud" at the beginning was used to name the first school reconstruction project during 2009–2010, donated by an alumnus from New Asia College of the Chinese University of Hong Kong. Since the pronunciation of "Bud" in Chinese is the same as "Asia", the donor and the architect named it. It also refers to the English word "bud", implying that small segments are assembled into one architecture as plants develop from buds.

"新芽"原是用来命名2009–2010年间的两个学校重建项目的，这两个项目由香港中文大学新亚学院的校友捐赠。由于"芽"与"亚"的中文发音相似，所以捐助者和建筑师用"新芽"来命名它。同时也取英文"bud"（芽）的意思来寓意这个组合系统，就像嫩芽成长为植株一样，小片段可组成一个建筑。

Credits and Data

Project title: New Bud Study Hall in Dazu Primary School
Project type: School
Location: Yanyuan, Sichuan, China
Completed: 2010
Principal architect: Nelson Tam Sin Lung, Zhu Jingxiang
Design team: Xia Heng, Zhang Dongguang, Gu Tian
Structure: Huang Shiping, Zhu Jingxiang
Size: 260 m^2

Project title: Work Station in Anzihe Panda Nature Reserve
Project type: Office + Exhibition
Location: Chongzhou, Sichuan, China
Completed: 2011
Principal architect: Zhu Jingxiang, Xia Heng
Design team: Zhang Dongguang, Wu Chenghui
Structure: Zhu Jingxiang, Wu Jing
Size: 200 m^2

Project title: Work Station in Baishuihe Panda Nature Reserve, Sichuan
Project type: Office + Exhibition
Location: Xiaoyudong, Baishuihe, Sichuan, China
Completed: 2013
Principal architect: Zhu Jingxiang, Zhang Dongguang
Design team: Xia Heng, Han Guori
Structure: Wu Jing, Zhu Jingxiang, Li Jing
Size: 140 m^2

Project title: Grameen Bank of the Yunus Centre
Location: Xuzhou, Jiangsu, China
Project type: Bank + Community Centre
Completed: 2014
Principal architect: Zhu Jingxiang, Han Guori, Xia Heng
Design team: Zhang Dongguang, Huang Zhengli, Wu Chenghui, Zhao Yan
Structure: Li Jing, Huang Shiping, Wu Jing
Size: 240 m^2

***Work Station in Baishuihe Panda Nature Reserve, Sichuan**, completed in 2013, was commissioned by World Wildlife Fund (WWF). It was the fifth application of NBS. The station is situated on a hill slope in the entrance area of a nature reserve. The project intended to demonstrate making use of the landscape instead of flat rice fields in reconstruction. A monorail transportation system was first set up to transport materials when construction started. It was used to deliver supplies and visitors after occupation. The two-story building is elevated, sitting on tubular piles. The pitched roof creates a tall and straight look and makes the second floor spacious and bright. Photos by Zhang Dongguang, courtesy of the architects.*

四川白水河自然保护区工作站，完成于2013年，是受世界自然基金会（WWF）委托的一个项目，为新芽建筑系统的第五次应用。该保护站坐落在自然保护区入口区域的一个小山坡上。项目旨在高效利用山坡地进行重建，而不是占用平坦的稻田地。山地缆车系统在建设之初便被用来构建短距离运输，房屋建设完成后又被用来运送物资和游客。两层楼高的建筑坐在管桩上被托举起，并以斜屋顶创建一个挺拔的外观和宽敞明亮的二楼空间。

New Bud Study Hall in Dazu Primary School*, completed in 2010, is a compact single story building as the second application of NBS, housing three classrooms and a reading lounge, without wasting the precious 260 m² space on corridors. Each room has distinct proportions and orientation in order to give kids from ethnic minorities a clear sense of location when inside. A simple timber-trellis cladding design allows the Study Hall to fit into the vernacular of the village while reducing the consumption of the logs required in building such a space with traditional methods. Photos by XIA Heng, courtesy of the architect.*

新芽达祖小学是一个紧凑的单层建筑，建于2010年，是新芽建筑系统的第二次应用，占地仅260m²，设有3间教室和1个阅读空间，采用了紧凑的扇形平面布局，没有在廊上浪费宝贵的可用面积。室内4个空间的尺寸、比例与朝向各不相同，使少数民族的孩在室内可以自然地感知所处的课室。简单的外围木制格架设计，使这个现代的小学融乡村氛围，又毋须像当地传统的井干式建筑般消耗许多原木。

Work Station in Anzihe Panda Nature Reserve *was completed in 2011 at the end of a road, standing in a forest with tall fir trees. The 11 m tall building is conceived as a cage structure and has three stories lifted up by pad foundations. Infill panels and diagonal tension rods provide additional support against buckling for slender columns. Room height, window level and stair position vary from floor to floor, differentiating the experience and hinting at the appropriate function. In order to ensure a smooth diagonal loading path, bay windows with polyhedral form offering rigidity were distributed on facades in a homogeneous chessboard pattern. Photos by Xia Heng, courtesy of the architect.*

崇州鞍子河大熊猫自然保护区工作站，建于2011年，它坐落在进山公路的尽头，在高大的杉树林中若隐若现。建筑整体架空于点式基础之上，形成整体的笼状结构。这栋11m高的建筑共有3层。填充板材与斜向拉杆使细长的结构立柱免于屈曲。每层房间的高度、窗位和楼梯的位置各不相同，形成差异化的体验，并暗示相应的功能。多面体的窗洞使用匀质的棋盘图案分布在立面上，在强化自身结构的同时，也确保了结构的对角传力路径。

自 2008 年以来，香港中文大学朱竞翔教授和他的团队开发了一系列轻量建筑系统，这些系统具有卓越性能，其质量令人赞叹。“新芽”建筑系统便是其中之一，它是由轻钢龙骨与多层木基板材加强组成的复合结构。通过这类系统建造的房屋不仅重量轻，而且具备良好的抗震能力。多层次的表面构造连续包裹建筑，可提供全面的隔热保护，提高室内的舒适度。在现场搭建仅需 2–4 周的干法施工，对周边环境的影响非常小。这类系统适合 4 层或以下的地表建筑，由这种系统所建造的建筑可实现易拆卸、易搬迁的可能，也易于实现外墙的个性化定制。

四川省“5 · 12”汶川大地震发生之后，来自香港中文大学的设计团队将“新芽”系统多次应用于内地地震多发地区。应对偏远地区资源匮乏的挑战，设计师们研发出了灵活的设计策略和有效的工具，通过学术研究将内地乡村与工业现实连接起来，展现出一系列兼顾现代性与本地特征，并回应可持续发展问题的解决方案。

本篇中的项目的所有预制组件都在邻近的成都和崇州工厂进行生产，以减少运输花费并结合利用当地资源。这种策略保证了建筑在严苛的预算条件下的优良品质。

在上述一系列项目中，团队不仅负责设计与建造，也承担了从早期预算规划到后期性能检测之间的许多其他任务。通过这些项目，该系统已证明了其广泛的应用领域，而团队则创造出了一种对用户和当地建筑行业而言，非常独特的可持续发展方式。这些建筑恢复了建筑师在社会中的原始角色职能——建设并守护社会。

***ameen Bank of the Yunus Centre** at Lukou of Xuzhou was a project CUHK
ım was invited to design. From design to production and construction, the
›ject took only eight weeks. Component assembling for the main body and
situ construction took four weeks. The team adopts close-range distributed
›duction to economize logistics. Most builders were local villagers, and bricks
ginally in there were reused. Photos by Han Guori and Xia Heng, courtesy of
› architect.*

徐州陆口尤努斯中心格莱珉银行，是香港中文大学团队于2014年11月应邀设计的一个项目。项目从开始设计到生产建设完成，仅用了8周时间。主体的组装和现场搭建用时4周。团队利用在邻近工厂进行分布式生产的方式，节约了物流。在这个项目中，大多数的工匠是当地的村民，原有房屋的砖头也得到了回收使用。

›nstruction of Grameen Bank of the Yunus Centre
›努斯中心格莱珉银行的施工现场

Mock-up of decorative high-pressure laminated panels
高压装饰层积板大样

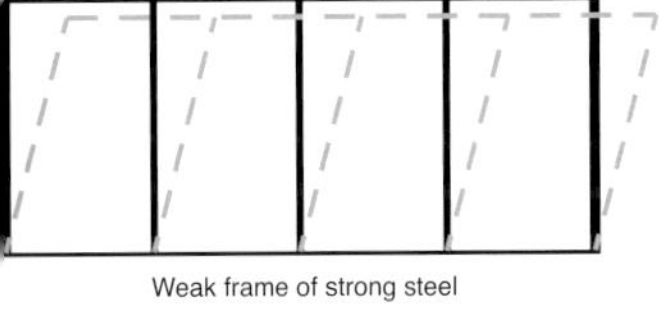

Weak frame of strong steel

Strong wall of weak infill panel

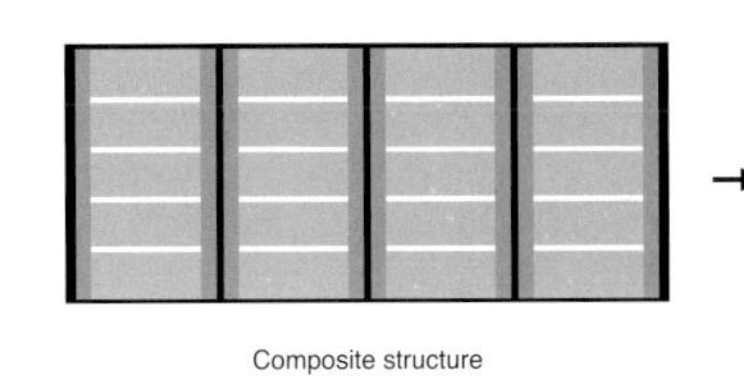

Composite structure

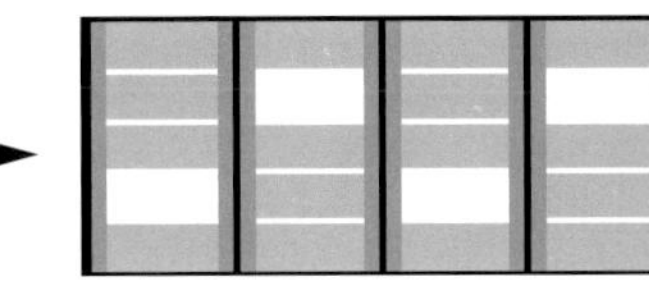

Careful opening distribution

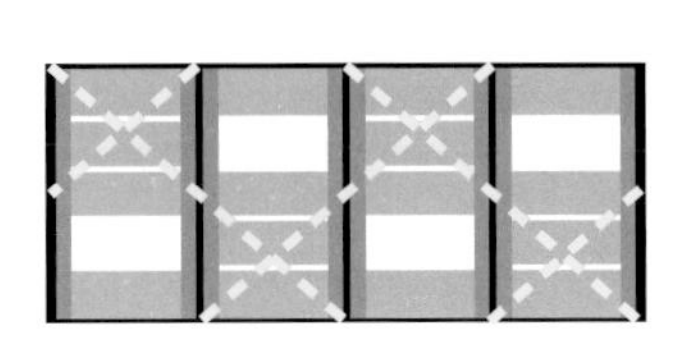

Hidden path of internal force

›iagram of NBS composition / 新芽建筑系统的构成图

OPEN Architecture

HEX-SYS

Guangzhou, Guangdong, China 2015

OPEN 建筑事务所

六边体系

中国，广东省，广州市 2015

HEX-SYS is a reconfigurable and reusable building system OPEN Architecture designed. As our reaction to the unique Chinese phenomenon in the recent decades' building frenzy – the production of vast amount of flamboyant but short-lived buildings, this modular building system can easily adapt to many different functions, and can be disassembled and reused, thus extending a building's life cycle and preventing huge waste of resources. Also by being modular and largely prefabricated, it can be built much faster than traditional buildings. This building system is part of OPEN's continuous efforts in exploring the ultimate potential of building sustainability.

Inspired by both the ancient Chinese wooden building system which can be taken apart and rebuilt elsewhere with little damage, and Le Corbusier's Swiss Pavilion which summarized his lifelong research on modular building systems, we designed this prototype comprised of hexagonal cells with architectural, structural and mechanical systems all synthesized within the same geometrical rules. The composition of cells can be rearranged according to different site and programmatic needs.

The basic building cell is a 40sqm hexagon module, with an inverted umbrella structure standing on a single pipe column which double functions as the rain flue. Rainwater is collected and used for landscape irrigation. There are 3 basic types of cells, indoor-open, indoor-closed and outdoor-open, to accommodate different functional needs. A 'missing' hexagon in a cluster of cells forms an internal Zen garden, like the void in Chinese paintings. The exterior façade is made of unitized curtain wall system. In order to maximize the recyclability and reusability of the building components, all connection details are designed to be reversible, no welding or glue allowed.

The structure sits on top of independent piles, floating above the existing parkland, while the small amount of excavated soil is use in landscaping to form small mounds, which frames an open spac for public events. Sandblasted and anodized aluminum panels are used for exterior cladding for their durability and low maintenanc Bamboo, the rapidly renewable material, is used throughout the interior spaces wherever wood is needed. The first realized prototype of HEX-SYS is in Guangzhou China, next to the city's new train station. It functions as an exhibition pavilion containing open display areas, offices, multimedia room, lounge and café.

pp. 132–133: View from the west. Rainwater is collected by roofs with an inverted umbrella structure. This page: View from indoor-open cell to the outdoor. All photos on pp. 132–135 by Zhang Chao, courtesy of the architects.

132–133页：从西侧看建筑。通过屋顶倒的伞状结构收集雨水。本页：从室内透单元看向室外。

Credits and Data

Project title: HEX-SYS
Client: Vanke Guangzhou
Program: Reception, exhibition, lounge, multi-media room, café, office
Location: Guangzhou, Guangdong, China
Design year: 2014–2015
Status: Completed
Design firm: OPEN Architecture
Principals in charge: Li Hu, Huang Wenjing
Project team: Zhao Yao, Andrea Antonucci, Laurence Chan, Hu Boji, Thomas Batzenschlager, Zhang Chang
Local design institute: CABR Technology Co., Ltd.
Landscape design: Guangzhou Shangwo Landscape Design Co., Ltd.
Landscape contractor: Shenzhen Wenke Landscape Co., Ltd.
Building area: 680 m²
Site area: 5,680 m²

六边体系是 OPEN 建筑事务所研发的灵活可拆装建筑体系。作为对中国近年来伴随建造热潮而出现的大量临时建筑的回应，这个可快速建造和重复使用的建筑体系延长了建筑的生命周期，实现了真正意义上的可持续性。预制化生产和装配式建造，使其像产品一样具备批量生产的可能。而通过模块的不同组合方式，它又会演化出各种各样的版本，灵活适用于不同的场地和功能。

设计的灵感源于可拆卸重组的中国古代木构建筑体系和勒·柯布西耶的瑞士展馆（融汇了他毕生研究模数化建造系统的成果）。这套系统旨在将结构、机电、外围护和室内装修等全部建造体系整合到可灵活拼接的六边形基本单元中，在严谨的几何规则的控制下，可自由地拼接组合各单元。

基本建筑单元是一个 40m² 的六边形模块。倒伞状的屋顶钢结构由位于中央的圆柱结构支撑，空心的圆柱兼作雨水管，可将收集到的雨水用于景观灌溉或注满庭院水池。3 种不同的单元——透明的、围合的、室外的，分别适应不同的功能需求。在一组单元中“缺失”的一个六边形，被设计为内部庭院，成为这个工业化建造体系里具有禅意的“留白”。建筑外围护是单元式幕墙。为了最大限度地实现建筑构件的可回收和可重复使用，所有的连接节点都不是焊接或打胶设计，以便拆卸。

建筑结构为独立桩基，可轻盈地漂浮于公园绿地之上，不对原有场地造成任何破坏。挖出的少量土方在主体建筑边堆成一个小山丘，围合出一个开放的公共活动空间。喷砂阳极电镀铝板则因其耐久性和易维护性用于所有的外墙覆盖层。速生的竹子作为木材的替代，是室内空间的主要材料之一。第一个六边体系原型已建成，位于广州南站附近，是一个包含办公、洽谈和咖啡区功能的展示中心。

This page: View of courtyard with a pond. A "missing" hexagon in a cluster of cells forms an internal Zen garden. A micro-climate is created in this courtyard.

本页：水池中的中庭。“缺失”的六边形单元，形成一个禅意庭园，营造出一种中庭的微气候。

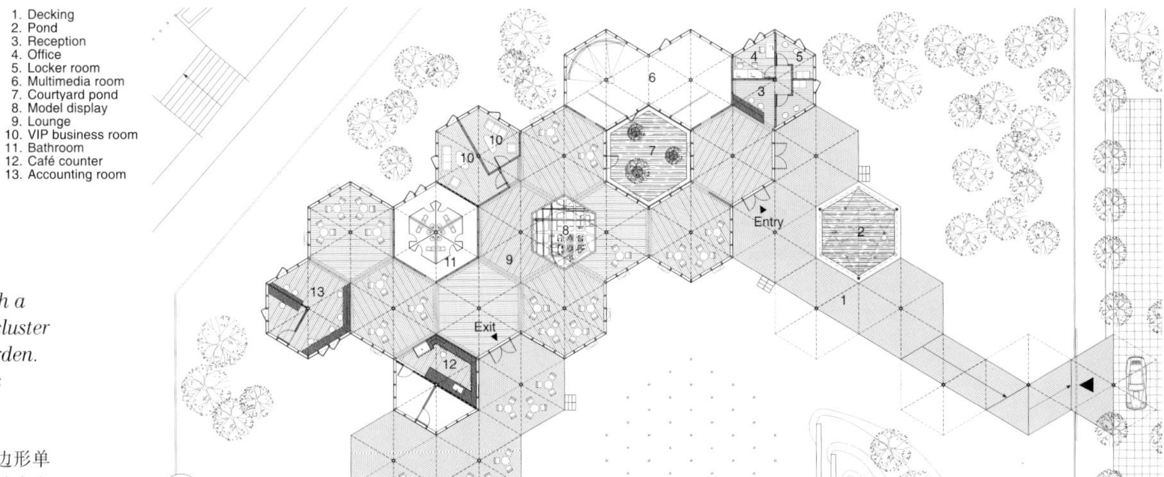

Plan (scale: 1/600) / 平面图（比例：1/600）

Tao Lei Architecture Design (TAOA)

The Concave House

Benxi, Liaoning Province, China 2010

陶磊建筑事务所

凹舍

中国，辽宁省，本溪市 2010

Design Background

This project, including multiple functions such as residence, studio and gallery, is specially designed for Dazhong Feng, who is a prestigious artist in China. The site is located in the downtown area of Benxi, Liaoning Province, facing mountains which turn into the best landscape view for this project. With the aim of relieving the conflicts between the fast pace of an urban lifestyle and the peaceful mind of the artist, this project is intended to create a harmonious internal space and interaction between nature and architecture. The building is constructed neither for public nor purely private purposes. Therefore, it not only requires a quiet living space for the artist but also a dynamic gallery space for visitors. As a whole the building is filled with profound cultural experience.

Design Concept

1) The AO (凹) shape

The building is inspired by the Chinese character 凹 , which means concave. The shape is a cube with all side roofs inclining downwardly to the centre. Three interconnected yards create the atmosphere that all views are pulled into the interior of the building. This concept also coincidentally matches a traditional Chinese housing design mode. The roofs are designed to block out the view of surrounding buildings so that when people stand at the platform in the middle of the rooftop, they can only feel the sky, the moonlight, themselves, and the time passing by.

2) The inner courtyards

Within the brick cube, the artistic creation of the book yard, the bamboo yard, and the mountain yard has made the interior space unique and full of surprise while the exterior is stable and serious. These yards are like Chinese lanterns lightening the rooms surrounding them and creating a dramatic atmosphere. This design origins from traditional Chinese building modes and culture but using a modern way of expression and regional criticism to reconsider traditions, which lead to the result of an oriental modern architecture.

3) The brick skin

Since the site is located in the northeastern China, thermal insulation becomes crucial. A custom-made 600 mm long brick can solve this problem and the color makes people feel warm. Being different from the roughness and hard character of the single brick, the skin is specially designed as a translucent coating which has openings that grew out and faded away. It forms into a network of tension and blurs the boundary between interior and exterior.

4) The twin walls

The twin walls create the entrance of the building, also acting as the courtyard walls that clarify the boundary. The inner wal which is parallel to the building uses the same design method the translucent building skins. The outer wall is solid and draw the boundary line of the site. These twin walls form a triangle water garden pointing to the entrance and the concrete entran extends out from the walls creating a sculpture-like space.

Opposite: View of the roof from above. The inward slanted roof blocks the view when standing on the roof platform. All photos on pp. 136–139 courtesy of the architect.

对页：从上方看屋顶。向内倾斜的屋顶挡了屋顶露台处的景观。

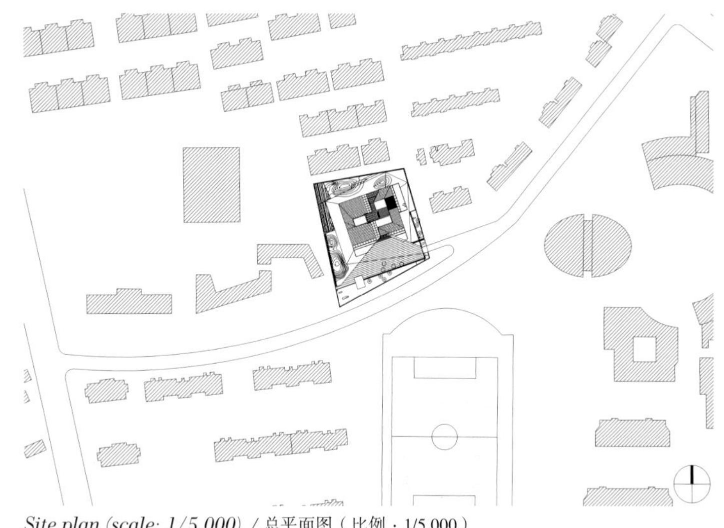

Site plan (scale: 1/5,000) / 总平面图（比例：1/5,000）

Tao Lei Architecture Design (TAOA)
The Concave House
Benxi, Liaoning Province, China

设计背景

这是为著名艺术家冯大中先生设计的住宅、工作室兼美术馆。项目位于辽宁省本溪市的主城区，正前方能遥望到山体形成的天然景观。设计旨在通过创造一个静态的内部世界，再由内而外展开与外部自然的对话，来应对快速发展的都市节奏与人文内心之间的冲突。它既不是公共建筑，也不是完全意义上的私人空间。因此，这就要求该建筑既要提供相对安静的创作与居住空间，还需一个可供来访者参观的画廊，建筑整体渗透一种恬静而深邃的文化感。

设计构思

1）凹形屋面

设计的灵感来自汉字“凹”。建筑整体呈一个内凹的方形“砖盒子”，屋面向中心汇聚，与 3 个庭院连接，巨大的空间张力将天空全部收纳到建筑内部，并暗合了传统的“四水归堂”。在屋顶的中心设置了可上人的木质屋面，由于凹形屋顶对周边城市的屏蔽作用，这里形成了巨大的场所感，在此仅能观望到远山、天空和明月，更易于感受四季的轮回和自己的存在。

2）屋中院

这个方形“砖盒子”的内部空间，通过书院、竹院、山院的插入，形成了“屋中院”，建筑整体成为一个外部严谨厚重而内部灵动的独立世界。插入的内院像灯笼一样点亮了整个室内空间，给建筑带来了无限的戏剧性。这是在中国传统的空间意识、文化意识及当下价值观的前提下去改变一些规则，营造的一个东方式的内部空间。

3）砖表皮

结合东北地域特点——寒冷，设计师为该建筑定做了色彩温暖且具有良好保温性能的 600mm 大砖，并试图让这种厚重且粗犷的材料呈现出与其原有属性相反的效果。将砖像拉伸的网眼织物结构一样进行垒砌，放眼到整体便形成了建筑的不透明到透明的渐变，从而获得了新的质感与张力。这种渐变模糊了室内和室外的界限。

4）双层墙

外院入口处是双层院墙，外实内虚。外墙的形状由用地决定，内墙为“漏”墙，与建筑平行。渐变镂空的内墙再一次契合了传统漏窗的功能和空间意境，不同的是创造了新的空间质感。在入口处，双层院墙交汇，被清水混凝土顶盖整合成相对完整且具有雕塑感的三角形水庭视觉空间。

This page: View of the entrance to the site . The water garden is seen to the left. Opposite: View of the book yard from the exhibition hall on the second floor.

本页：基地入口。左边是水庭空间。
对页：从二层展厅看书院中庭。

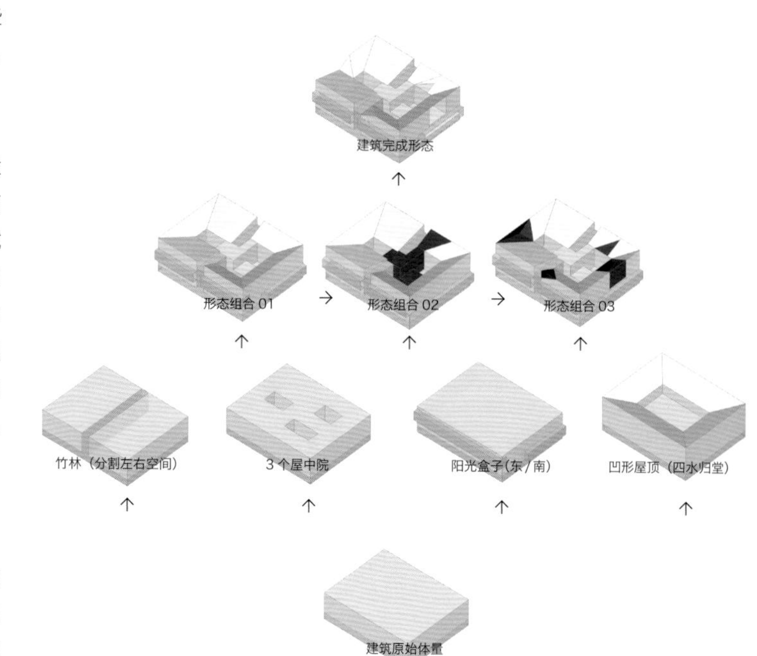

Concept diagram / 概念分析图

1. Atelier 1
2. Bedroom
3. Stockroom for pictures
4. Study room
5. Stockroom for books
6. Stockroom for paper
7. Exhibition hall
8. Atelier 2
9. Living room
10. Main bedroom
11. Central hall
12. Bathroom
13. Book yard
14. Bamboo yard
15. Mountain yard
16. Platform
17. Guest room
18. Monitor room
19. Leisure room
20. Main dining room
21. Kitchen
22. Domestic dining room
23. Main entrance
24. Side entrance
25. Door
26. Parking
27. Pool yard
28. Slope
29. Stockroom
30. Event space

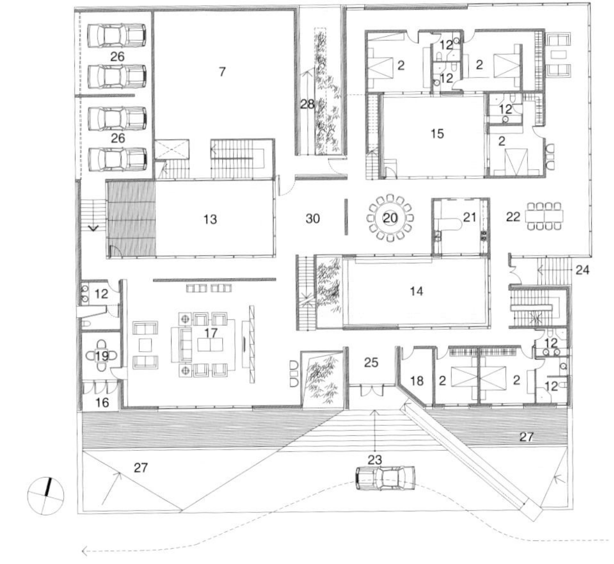

Ground floor plan (scale: 1/600) / 首层平面图（比例：1/600）

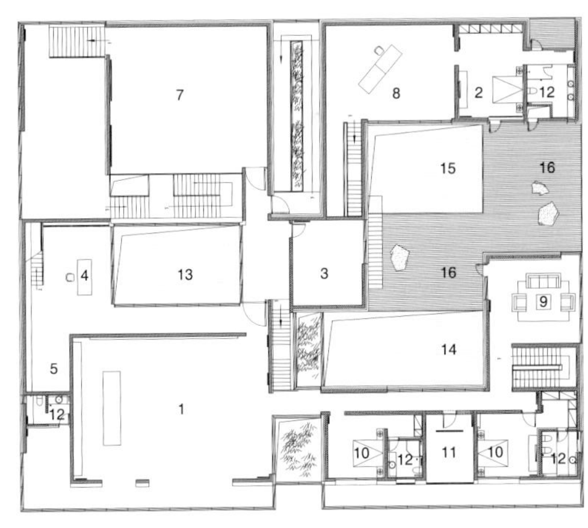

Second floor plan / 二层平面图

Credits and Data

Project title: The Concave House
Program: Residence, studio, private art gallery
Design content: Building, interior, landscape
Location: Benxi, Liaoning Province, China
Completion year: 2010
Architect: Tao Lei Architecture Design
Land area: 5,000 m^2
Structure area: 3,000 m^2

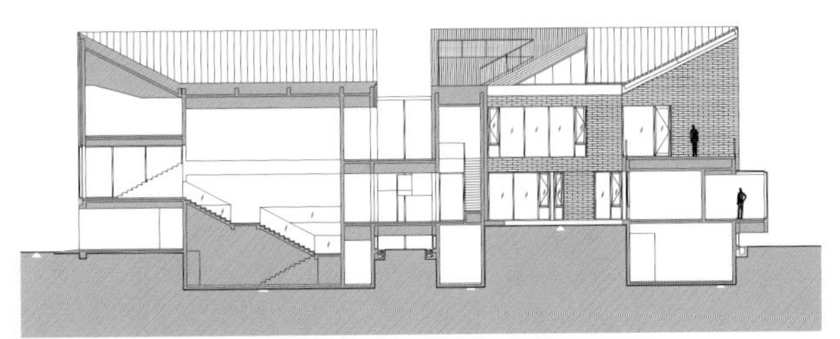

Section (scale: 1/600) / 剖面图（比例：1/600）

Natural Build Operation LLC

1178 Waima Road Warehouse Renovation

Huangpu, Shanghai, China 2014

直造建筑事务所
外马路 1178 号创意办公改造
中国，上海市，黄浦区 2014

Located at Shanghai's South Bund historic dockyards, the co-work loft was converted from part of a 1930s warehouse building previously owned by Huang Jinrong, once a tycoon of the notorious Green Gang.

The original space offers a pitched high ceiling up to almost 10 m under two parallel gables. Each gable spans 12 m, featuring a series of wood and steel combination trusses. The renovation approach is driven by the idea that the existing components should be well exhibited and reinforced in their new spatial context where functionality and flexibility are instilled to accommodate a shifting workforce and a diverse program of events.

Without taking any disruptive action, we approach this setting with a two-pronged strategy. First, the original L-shaped plan was reconfigured into a dumbbell, with one end facing the river serves as an event space, whereas an open-plan workspace takes up the other end. In-between, a corridor runs through the whole space, programmatically connecting the front and the rear. Lining this corridor are enclosed offices and a glassed conference room. Secondly, as the corridor unfolds, various forms of suspended ceiling have been explored in each cross section, on one hand to strategically keep all the MEP equipment behind the scenes, on the other, to create divergent patterns of light and shadow that vary with time of day and seasons of the year.

The corridor also employs the Venturi Effect on purpose, where the wind velocity increases as the cross sectional area decreases. In the summer, the breeze from the river is converged at the corridor and can reach all the way down to the workspace, meanwhile the high ceiling allows the heat to accumulate at the top and ventilated through the clerestories, thus providing convective cooling.

In another dimension, the use of add-on materials has been minimized. The ceiling's worn finish is left as it was. The decaying timber posts and corbels have been gently cleaned and treated. The original wide plank flooring is kept and repaired despite its irregular surface, which has been resolved by attaching levelers to the bespoke desks. All walls are finished in whitewash, not as homage to high modernism, but more like the museum walls that serve as a muted backdrop, to bring the weathered, unfinished floor and ceiling to the foreground.
As a whole, the design respects and celebrates the weathering of architecture – the constant refinishing of the building by natural forces which makes the final state of the building necessarily indefinite. In such inexpensive and low-tech environment, we try to create a raw and austere aesthetic not dissimilar to the original warehouse.

pp. 140–141: View of the workspace on the northwest side. This page: View of the corridor flanked by various spaces. All photos on pp. 140–143 by Hao Chen, courtesy of the architect.

140–141页：西北侧的工作室。本页：与各个空间相接的走廊。

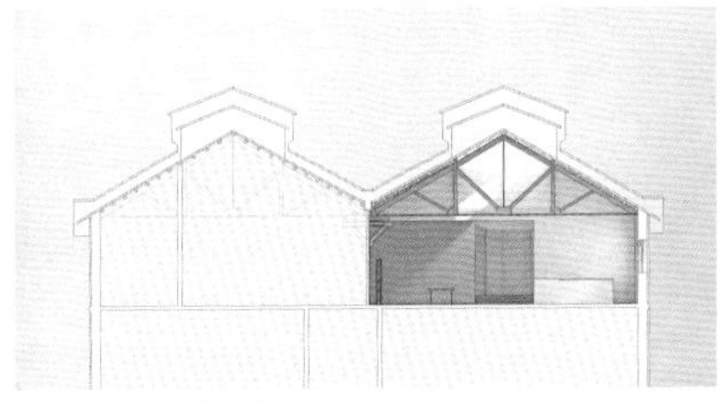

Section 1 / 剖面图 1

Section 4 / 剖面图 4

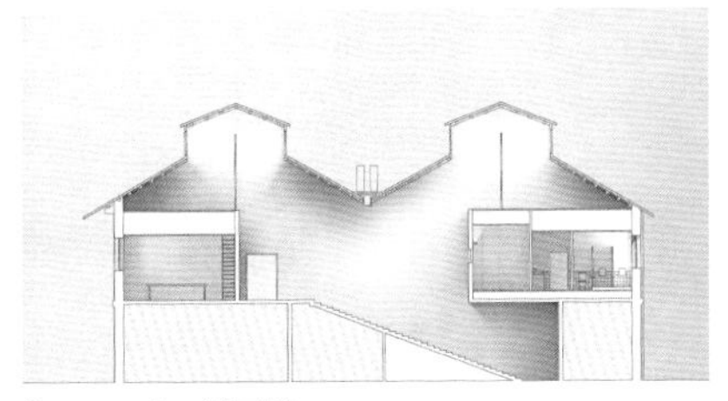

Section 6 / 剖面图 6

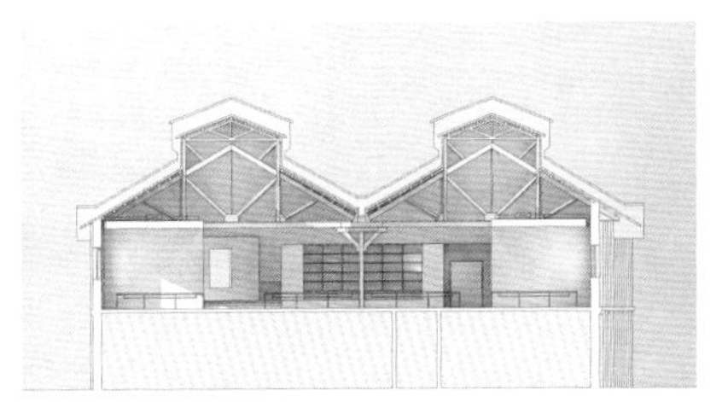

Section 8 / 剖面图 8

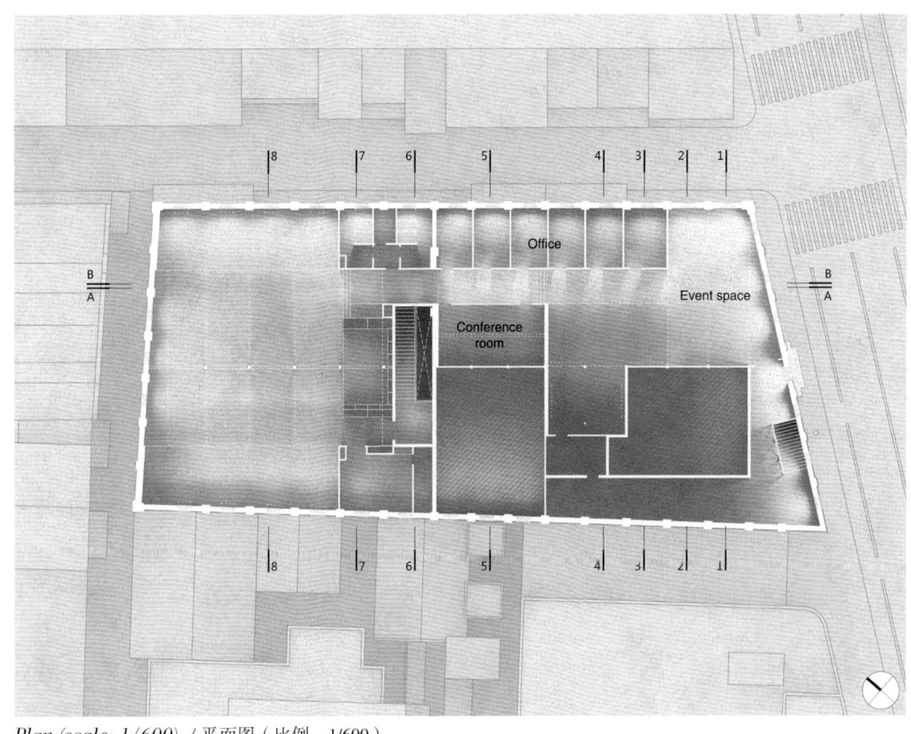

Plan (scale: 1/600) / 平面图（比例：1/600）

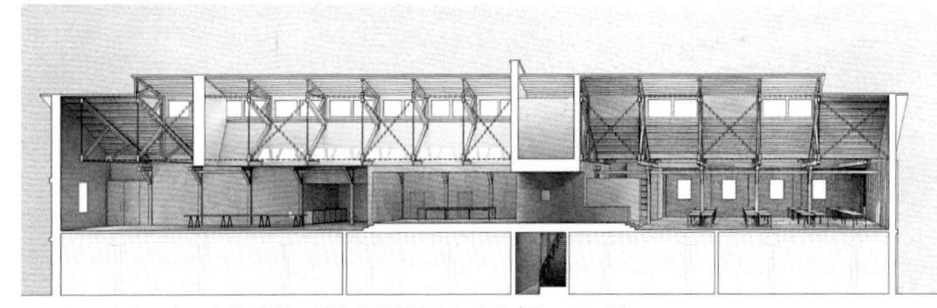

Section perspective A (scale: 1/600) / 剖面透视图 A（比例：1/600）

该项目位于上海南外滩货运码头区。20 世纪 30 年代，它曾属青帮大亨黄金荣的"新昌仓库"。

原始空间屋架的纵向尺度极大，由 2 个近 10m 高的平行双坡构成。屋架的跨度为 12m，由一系列的钢木混合木桁架组成。改造意图是在新的空间场景中自然而然地强化原始留存的构件。基于此，设计将功能的灵活性注入到这个空间，为多变的工作模式和活动提供可能。

在尽可能少的操作同时，设计师进行了两方面的设计策略。首先，原始"L"形的平面被重新组织为哑铃状，朝向江景且靠近入口楼梯的一侧作为活动空间，后侧则为开敞的办公区。连接公共区域和工作区域的是一条走廊。走廊两侧分别是一些小办公室及一个全透明的会议室。其次，在剖面上随着走廊的展开，不同形式的吊顶被用于各个部分。一方面，策略性地将机电系统隐藏起来。另一方面，光影关系随不同的时节更替而变换。

同时，走廊还应用了文丘里效应，即风速会随着横截面的减小而增加。夏季，来自公共区域窗外的江风在走廊处汇集，并一直吹入向后敞开的工作区域。走廊上方特意保留的高度和排气扇，使热空气上升后就可以立刻排到室外，从而实现对流降温。这类似于上海石库门弄堂里的穿堂风。

材质方面，尽可能地避免附加材料。设计师保留了表面磨损的屋面板，并轻柔地清洗处理了逐渐风化的木柱和枕梁，同时修复了原始的宽木地板，包括它不规则的表面，通过定制办公桌的矫直机将其柔化。设计还将所有的墙面刷为白色，这不是对现代主义的致敬，而是类似博物馆里的展墙，让屋顶和地面成为前景。

总体上，设计尊重并颂扬了自然对建筑的侵蚀性——在自然力的作用下，建筑的体表得到不断的更新，从而获得一个无法确定的最终状态。通过这样一种平价且低技的手法，设计试图传达出旧物本身带给场所的静谧感和一种简朴甚至清苦的环境美学。

Credits and Data

Project title: 1178 Waima Road Warehouse Renovation
Location: Huangpu, Shanghai, China
Architects: Natural Build Operation LLC
Project year: 2014
Project area: 960 m²

TM Studio

Han Tianheng Art Museum
Jiading District, Shanghai, China 2013

童明工作室
韩天衡美术馆
中国，上海市，嘉定区 2013

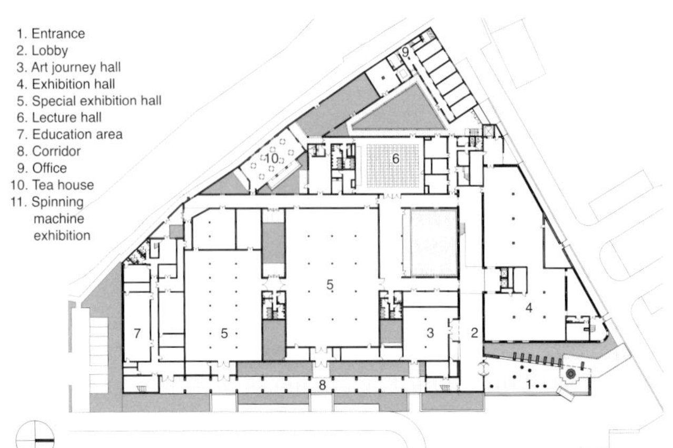

Ground floor plan (scale: 1/2,000) / 首层平面图（比例：*1/2,000*）

pp. 144–145: Looking down to the tooth roofed part from the lobby on the third floor. Photo by Lv Hengzhong. All photos on pp. 144–147 courtesy of the architect.

144–145页：从三楼大厅俯瞰屋顶的锯齿状结构。

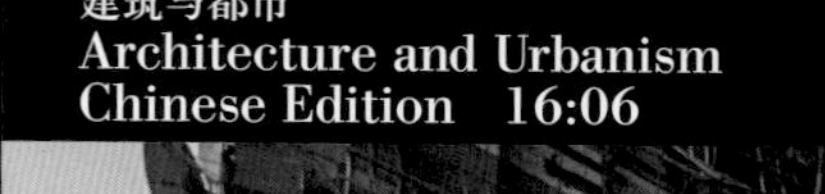

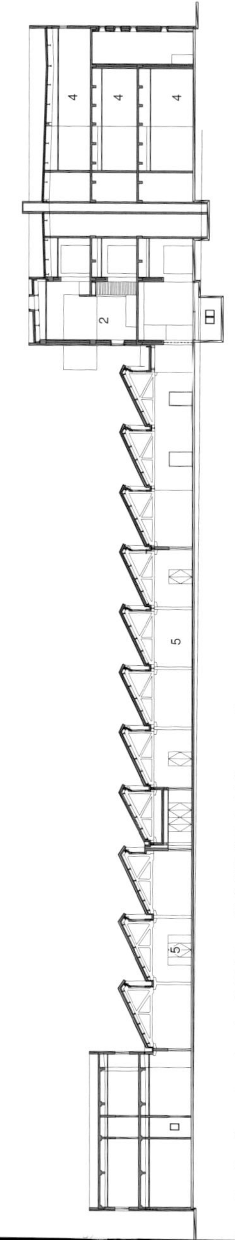

Longitudinal section (scale: 1/800) / 纵向剖面图（比例：1/800）

.n Tianheng Art Museum is a project to renovate the Feilian xtile Mill, which is preserved as an industrial heritage site Jia Ding, a satellite town northwest of Shanghai. The textile ill enjoyed a history of more than 70 years. It developed since e 1940s from a small, two-row workshop to a big enterprise cupying a site of two hectares.

vo parts can be distinguished in the general layout of this xtile mill: a pre-1980s traditional saw tooth roofed factory, d the surrounding masonry buildings as extensions. Due to a ocess of gradual expansion, the factory actually consisted of evious and chaotic parts and sections. Therefore, connections d relations among the different pieces of the program was a y component in the architectural design. The design proposes strengthen and give a clear differential value to the perimeter ternal facilities, in order to make the freeing of central space ossible , giving maximum flexibility to future adaptations of e functional program.

he most interesting part of the building is the one-story saw oth roofed factory, which has been successfully reinforced, d equipped with insulating glass and AV system inside. Several ardens were inserted within the galleries for public exhibitions. he red tile roof can be enjoyed as the fifth elevation by visitors om the surrounding higher buildings, some of which were organized to provide spaces for craft education and restoration orkshops as well as event and restaurant areas.

he northern part of the complex is a three-story-high factory uilding developed in 1993, occupying a major percentage f the floor area of the complex. It was transformed to ermanent galleries for hosting and displaying the collections of alligrapher, painter, connoisseur and collector Han Tianheng, ne of the most important figures in traditional Chinese art in hanghai.

n order to integrate all these parts, a new structure was esigned and placed between the original two parts as the obby. This new three-story high structure is in an L shape, the ntention of which is to bend the lobby inside out to the east, o as to accommodate a stage-like entrance to the museum.

'he public space in the lobby is a large, linearly arranged foyer djoined by the event area, auditorium, retail spaces, and afeteria on the ground floor. A sculpturesque steel staircase vas designed in the central area, traversing the three-story oid to achieve the vertical communication. The image for the efurbishment was conceived in the color black at the very beginning, in order to provide a contrast with the heritage parts, which were well preserved in their original appearance.

Working with the various historic elements, the project of Han Tianheng Art Museum has an object of coordinating and ntegrating the new components clearly to the original structure, from which to generate not only continuous and coherent spaces, but also a calm and carefully designed ambience for the exhibited art pieces.

Credits and Data

Project title: Han Tianheng Art Museum
Client: Shanghai Jiading of state-owned Assets Management Co., Ltd.
Location: 70 Bole Road, Jiading District, Shanghai, China
Year of design: 2012
Year of completion: 2013
Architects: TM Studio
Design team: Tong Ming, Huang Yi, Huang Xiaoying
Site area: 14,377 m²
Floor area: 11,433 m²

韩天衡美术馆的前身是坐落于上海市嘉定区（上海市西北区域的一个卫星城）的飞联纺织厂。作为一处重要的工业遗产，这座纺织工厂拥有 70 多年的历史，始建于 1940 年代末，从只有两排简易的厂房发展成为占地 2hm² 的大型企业。

纺织厂的总体格局可以分为两部分，其一是 20 世纪 80 年代之前经过历次扩建形成的 11 跨锯齿形单层厂房，其二则是周边后来增加的砖混结构厂房。由于整座厂房无论是在平面还是在空间中都呈现出一种混杂而多元的状况，因此，如何在不同要素、不同功能之间加强联系、达成协调就成为建筑设计首先需要面对的问题。设计目标就是强化周边的外部设施，为其赋予与众不同的价值，以释放中央空间，为未来的功能需求预留出最大的弹性空间。

锯齿形厂房是现状格局中最有特色的一部分，对原有结构进行钢结构加固，安装了植入玻璃、空调系统等设备，公共展览区的画廊中穿插布置了若干花园。游客们可以从周边高层建筑内观赏到建筑的第五立面——红瓦屋顶。周围的建筑，有的经部分改造后成为手工艺教学、修补工作坊，有的则被用作活动或餐饮空间。

基地北部三层楼高的车间建于 1993 年，占据整个建筑综合体很大的面积。它被改造成常设展厅，用以收藏并展示上海著名书法家、画家、鉴赏家和收藏家韩天衡先生的各类收藏品及作品。

为了促使美术馆的各部分能够形成一个整体，设计师在既有建筑与改造建筑之间插入了一个新结构作为公共大厅。新结构三层通高，呈“L”形，由东向外折出，形成一个具有舞台效果的通向博物馆的入口门廊。大厅是垂直向上的公共空间，与一层的活动区域、演讲厅、小卖部和咖啡厅相毗邻。中央区域设计了一个雕塑般的钢制楼梯，为三层楼高的空间营造了一种竖向的交流感。建筑更新的部分被设想为黑色，以便与遵照原貌保存完好的历史建筑形成对比。

通过与诸多历史要素进行协调性设计，韩天衡美术馆达成了预定的设计目标，将新元素（现代）融入到旧结构（传统）之中，构造了完整而连续的空间感受，也为藏品展示营造了一种平静细腻的氛围。

Opposite: Extention part. View of the three-story void. Photo by Chen Hao. This page, left: Exterior view of the entrance. This page, right: Interior view of the lecture hall. Two photos by Lv Hengzhong.

对页：增建部分。看向三层通高的空间。本页，左：入口外观。本页，右：演讲厅的内观。

Scenic Architecture Office

China Fortune Exhibition Centre

Xuhui District, Shanghai, China 2013

山水秀建筑事务所
华鑫展示中心
中国，上海市，徐汇区 2013

pp. 148–149: Exterior view from the southwest. The building volume is raised to the second floor and placed among the existing trees. p. 150, above: View of the pool from the entrance of the signing room in the west corner. p. 150, below: Interior view of the signing room in the north corner. All photos on pp. 148–151 by Shinkenchiku-sha.

148–149页：从西南侧看建筑外观。建筑的空间被抬高至二层并设置在已有树木的中间。150页，上：从西端的签约室入口看向水池。150页，下：北端的签约室的内景。

Feature:
Architects in China

Scenic Architecture Office
China Fortune Exhibition Centre
Xuhui District, Shanghai, China

ope that this building can enlighten us to think about mutual evance among human, nature and society.

e China Fortune office complex is located to the west of Guilin ad, with a green area at the south of its entry. This green area s six old camphor trees and opens to the urban main road. ese two facts become the starting point of the design, and nce to lead to two basic strategies of the exhibition centre. e is to elevate the main body of the building up to the second or in order to maximize the open green space on the ground; cond is to establish an intimate and interactive relationship th the six trees while protecting them on the site.

e completed building is composed of four independently spended structures that are linked by bridges. Ten combined el and concrete walls support the upper structures, which e all covered by reflective stainless steel panels. These panels ntain all the vertical ducts and reflect the surrounding green vironment. The walls are thus cleared up and help the spension effect of the upper volumes. A ground floor atrium enclosed by transparent glass between three structures. It troduces scenery and natural light by all-around transparency d a skylight, and creates spatial interaction between inside d outside.

hen approaching the second floor through the stair in the rium, a new spatial order is unfolded along the path. Four spended volumes that were realized by steel trusses stretch emselves horizontally with "Y" or "L" shapes among the old ees. Twisted and tensioned aluminum strips assemble the anslucent walls of these volumes, which present the truss ructure in an indistinct way and become containers and undaries of a series of interior and exterior spaces. When andering across these translucent walls, the visitor will ternatively encounter rooms, courtyards, bridges and different eneries guided by them. The branches and leafs of the trees averse the building freely and become touchable friends.

ere the structure, its adherent material, and the branches and afs of trees interweave together to present the atmosphere of ch space. It is under the organization of time (or path) that ese spaces (room and courtyard) realize an environmental perience where time and space interact. It is a work of ollaboration by both architecture and nature.

'e might never expect kind feedbacks from nature unless we eat nature in a kind and positive way. The architecture of the 1st century shall not only respond to human needs, but also ct as a positive media between human and environment. The ssential goal of future architecture is to establish balanced and ynamic relevance among human, nature and society.

redits and Data

roject title: China Fortune Exhibition Centre
lient: China Fortune Properties Group
ocation: Guilin Road & Yishan Road, Xuhui District, Shanghai, China
rogram: Exhibition + teahouses
esign: 2012
uilt: 2013
rchitects: Scenic Architecture Office
esign Team: Zhu Xiaofeng (Design Principal), Ding Penghua (Project Designer), Cai Mian, Yang Hong, Li Haoran, Du Shigang
tructural & MEP: Shanghai Greenland Building Steel Structure Ltd.
onstruction: Greenland Construction Group
laterials: Mirror finish stainless steel, twisted & tensioned aluminum strips, transparent and fritted glass, solid and perforated aluminum panels, gravel, water
tructure: SRC, bended steel truss
loors: 2
uilding area: 760 m²
ite area: 5,500 m²

希望通过这座建筑，启发我们思考人与自然、社会之间的关联。

华鑫办公集群位于上海市桂林路以西，其入口南侧是一块绿地。这块绿地面向城市干道的开放属性，以及其中的6棵大香樟树，成为设计的出发点，并由此确立了展示中心的两个基本策略：一，建筑主体抬高至二层，最大化地开放地面的绿化空间；二，保留6棵大树的同时，在建筑与树之间建立亲密的互动关系。

最终完成的建筑由4个独立的悬浮体串联而成。底层的10片混凝土墙支撑着上部结构，并收纳了所有垂直的设备管道，表面包覆的镜面不锈钢映射着外部的绿化环境，从而在消解自身的同时，凸显出地面层的开放和上部的悬浮感。4个单体围合成通高的室内中庭，透过四周悬挂的全透明玻璃以及顶部的天窗，引入外部的风景和自然光，使空间内外交融。

沿着中庭内的折梯抵达二层，会进入一种崭新的空间秩序。4个悬浮体的悬挑结构由钢桁架实现，它们在水平方向上以"Y"或"L"形的姿态在大树之间自由地伸展。由波纹扭拉的铝条构成的半透"粉墙"，以若隐若现的方式呈现了桁架的结构，并成为一系列室内外空间的容器和间隔。穿行在这些半透墙体内外，小屋、小院、小桥以及它们所接引的不同风景，将在漫步的路径上交替出现。大树的枝叶在建筑内外自由穿越，成为触手可及的亲密伙伴。

在这里，建筑的结构、材质与大树的枝干、树叶交织在一起，一起营造出一组纯净的室内外空间。这些空间（屋和院）在时间（路径）的组织下，共同实现了时空交汇的环境体验。这是一件由建筑和自然合作完成的作品。

如果我们以积极的方式善待自然，那么我们也会得到自然善意的回馈。21世纪的建筑不仅要回应人的需求，更要积极担当人与环境之间的媒介。未来建筑的根本目的将是在人、自然、社会三者之间建立平衡而又充满生机的关联。

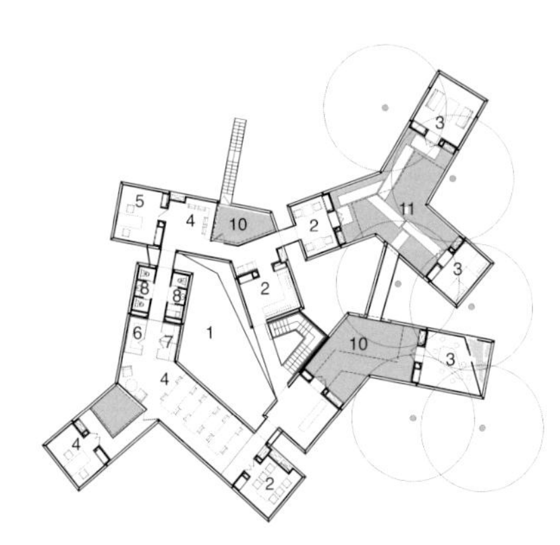

1. Exhibition room
2. Meeting room
3. Signing room
4. Office
5. Accounting room
6. Archives room
7. Changing room
8. Cleaning room
9. Reception
10. Courtyard
11. Pool

Second floor plan / 二层平面图

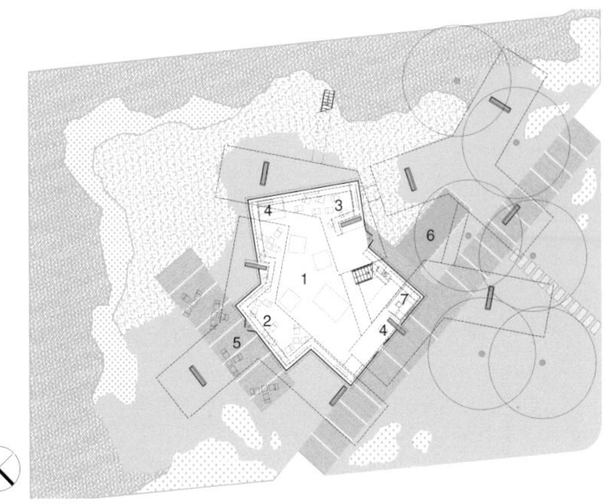

1. Exhibition room
2. Meeting room
3. Video room
4. Reception
5. Platform
6. Pool
7. Equipment room

Guilin Road

Ground floor plan (scale: 1/800) / 首层平面图（比例：1/800）

Atelier Archmixing

Facade Renovation for Building 8, Hengshanfang
Xuhui District, Shanghai, China 2014

阿科米星建筑设计事务所
衡山坊 8 号楼外立面改造
中国，上海市，徐汇区 2014

This facade renovation project is a small part of a big urban regeneration program, Hengshanfang, which intends to transform a historical residential district with Lilong housing (residentials in lane: specific typology in Shanghai) built in 1934 and villas built in 1948 into a boutique commercial area. The site is located in Xujiahui, one of the busiest commercial centres in Shanghai, where constant change plays a crucial role in guaranteeing economic and urban vitality. Thus, for Atelier Archmixing, the biggest challenge is how to achieve iconic difference as well as commercial dynamics without sacrificing the coherence of a historical preservation area.
The old villa stands in the centre of the whole block, with a corner exposed to the main entrance, the intersection of two leafy streets, as well as a large and popular urban park. In order to balance integration and distinction, the architects decided to cover the three-story structure with a changeable skin, which remains elegantly quiet during the day while sending out charming allure during the night. Instead of using a conventional curtain wall or floodlighting, they applied illuminated brick, an originally designed and customized product to achieve this dramatic effect. The historical building, now a boutique shop, is covered with an envelope combining two different materials – traditional gray bricks mixing with this unique illuminated bricks, both sharing the same size and similar colors. Therefore, during the daytime, with a pure brick texture, it easily integrates into its surroundings. When night falls, the wall suddenly becomes shining, just like lighting up a lantern with rich patterns, successfully distinguishing itself, and at the same time attracting and amusing the pedestrians. What makes the building stand out in this downtown commercial area is the dramatic fact that this is an everyday transformation of the building itself instead of a neon light effect. Moreover, this sophisticated design of difference, changing from daytime quietness to nighttime brilliance, also conveys a sense of Shanghai character – no matter whether in a mode of withdrawn seriousness or open enchantment, it always stays restrained and elegant.
The illuminated brick is a complicated product composed of five different segments. A stainless steel box in the size of a normal brick serves as the main structure, inserted with a lighting facility made of three different layers, among which a new material called Photosensitive Mineral Resin Sheet plays a key role in creating delicate texture. These bricks were laid alternatively with gray bricks in mortar, just like building a traditional brick wall. Four elevations constitute a continuous skin to achieve unity and simplicity.
The illuminated facade is composed of overlapping rectangular lighting patterns of varied size and intensity, intersecting with window frames. Since the LED bricks are of the same size as those gray ones, they can form rectangular lighting areas of different texture. Rectangular is also the shape of windows and doors – it is applied as the basic form. To exaggerate this covering effect and create different brightness patterns, small-sized and extremely dense lighting blocks are hidden in those large areas.
Both the doorway and shop windows project from the exterior walls to satisfy functional as well as aesthetic purposes. The entranceway is totally made of stainless steel plate. The same material is also applied to the prominent windows as exterior frame, but the interior surface is covered with sheet copper. While the stainless steel produces illusive reflections of the adjacent lighting bricks, the golden copper adds a charming sense of warmness and prosperity.

Opposite: View of the lane from the south. The illuminated bricks were designed for this facade renovation project, inspired by the iconic nightscape of Shanghai. All photos on pp. 153–155 by Tang Yu, courtesy of the architects.

对页：从南侧看甬道。发光砖是受到上海标性夜景的启发，为改造项目特别设计的。

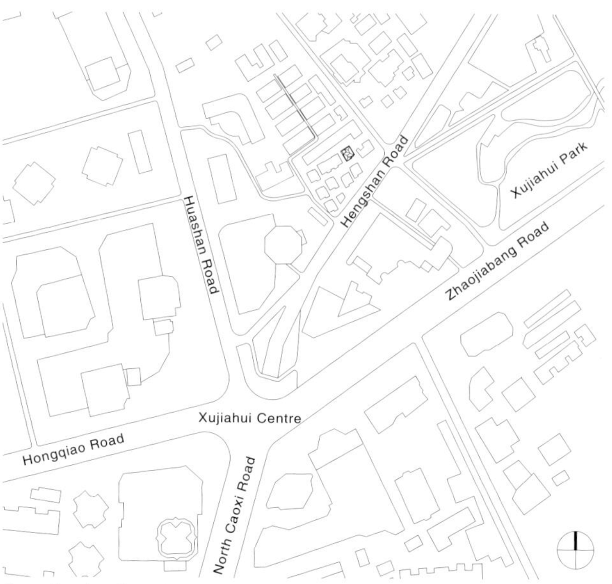

Site plan (scale: 1/8,000) / 总平面图（比例：1/8,000）

建筑与都市
Architecture and Urbanism
Chinese Edition 16:06

Feature:
Architects in China

Atelier Archmixing / Facade Renovation
for Building 8, Hengshanfang
Xuhui District, Shanghai, China

衡山坊历史保护区由建于 1934 年的里弄住宅和建于 1948 年的花园洋房构成，拟改造成一个精品商业区。衡山坊 8 号楼的立面改造就是这个大城市更新的一部分。基地所在的徐家汇商圈，是上海最繁华的商业中心之一，其经济和城市的活力来源于不断发生的变化。因此，这个项目对于阿科米星来说，最大的挑战是：如何在不牺牲历史保护街区统一性的情况下，实现新旧建筑之间标志性的差异，以及商业的活力。

这幢老洋房坐落在整个街区的中心，一角朝向商业区的主入口，紧邻衡山路天平路的十字路口，还有广受欢迎的徐家汇公园。为了平衡差异性和统一性，建筑师为这座变身为精品店的三层老洋房覆盖了一层可变的肌肤——白天优雅安静，夜晚魅力四射。这种戏剧性的效果由一种原创的发光砖产品而非传统的幕墙或照明灯实现，这种独特的发光砖与传统青砖有着相同的尺寸，并与青砖混合形成表皮。白天，它是纯粹的清水砖效果，很容易融入环境；夜幕降临时，墙面会突然闪亮起来，仿佛神奇地点亮一盏质感丰富的灯一样。这不仅使建筑从周围环境中脱颖而出，也吸引了路上的行人，为他们带来惊喜。然而，使这座建筑从中心商业圈内脱颖而出的并不是寻常的霓虹灯效果，而是建筑本身在一日之内不断变幻的戏剧感。这种建筑从白天的“静谧”向夜晚的“璀璨”过渡的精妙设计，也传达出一种“上海性格”：隐藏时的板正与放开时的漂亮都透着点儿矜持。

发光体组件由 5 个不同的部分组成：主体结构是与传统砖一样大小的不锈钢框，其上插入由 3 个层次构成的照明设施。其中拓彩岩透光板这种新型材料在创造精致纹理时起到了关键作用。发光砖和青砖按照传统的砌筑方式有序地铺砌，形成大小和密度各不相同的方形发光区。4 个立面构成了统一又简单的连续表皮。

发光的建筑表皮由相互叠合的方形发光图案组成，这些发光图案大小各异、密度不同，并与方形的门窗框相交叠。由于 LED 砖与其他青砖的形状大小一样，所以它们可以一起构成不同肌理的方形发光区域。方形也是建筑中门窗的形状，它在表皮中成为了一种基本元素。为了突出这种表皮覆盖的效果并生成不同亮度的肌理，在大的发光区域内还隐藏着一些小尺度、高密度、亮度特别高的发光区。

入口和橱窗突出于外墙，以满功能和审美的需要。入口由不锈钢构成，同样的材料也应用于独特的窗外壁，窗户内壁则覆盖了铜板。锈钢反射邻近的发光砖产生炫丽的影，亮铜色则为建筑增添了迷人的暖感和繁华感。

This page: Night view of the west elevation. Illuminated bricks surround the openings and shine at night.
Opposite, above: View of the entrance and openings made of stainless steel plates to reflect the illuminated bricks.
Opposite, below: Close-up view of illuminated bricks and window frame.

本页：西侧立面的夜景。发光砖围绕着口在夜晚闪耀。对页，上：不锈钢板做的入口和开口可以反射出发光砖的光亮对页，下：发光砖和窗框的近景。

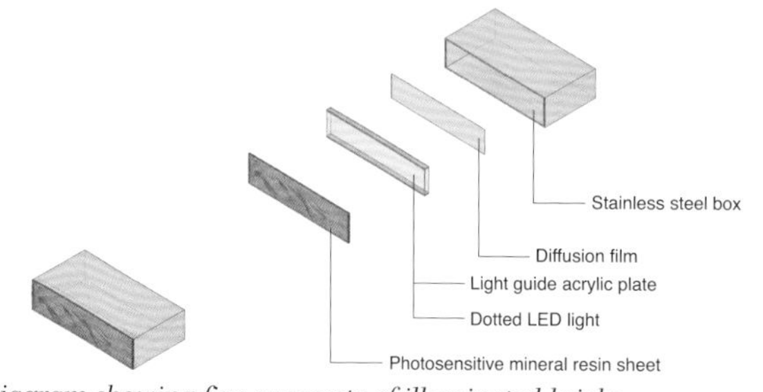

Diagram showing five segments of illuminated bricks
发光砖 5 个组成部分的图示

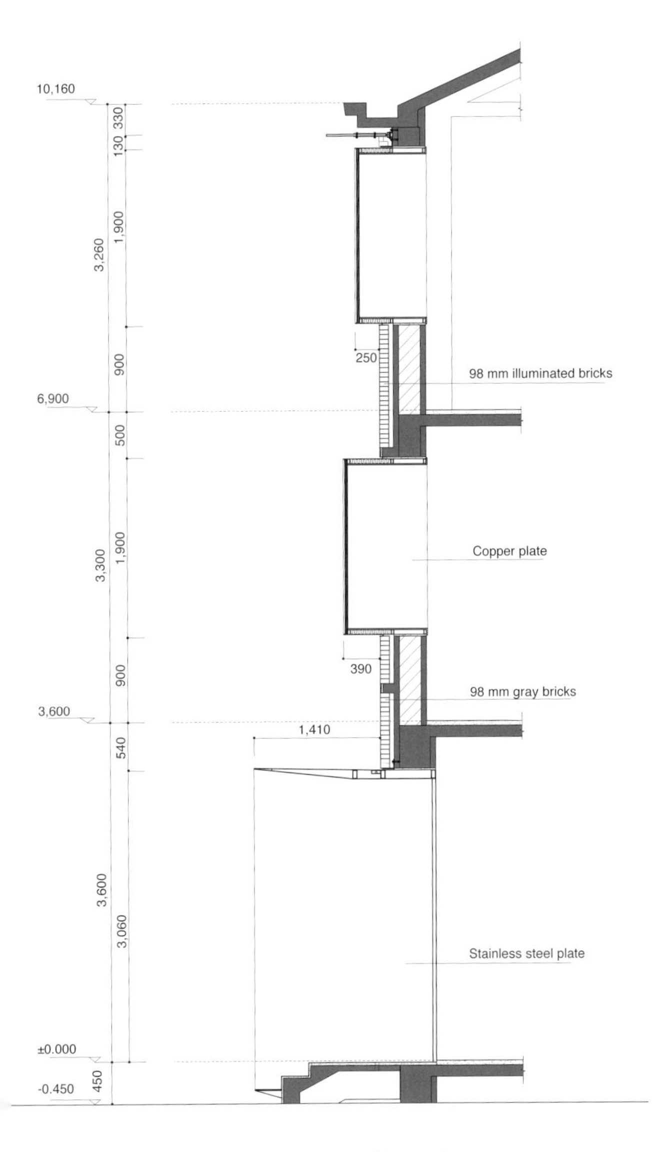

ction detail (scale: 1/80) / 剖面细节图（比例：1/80）

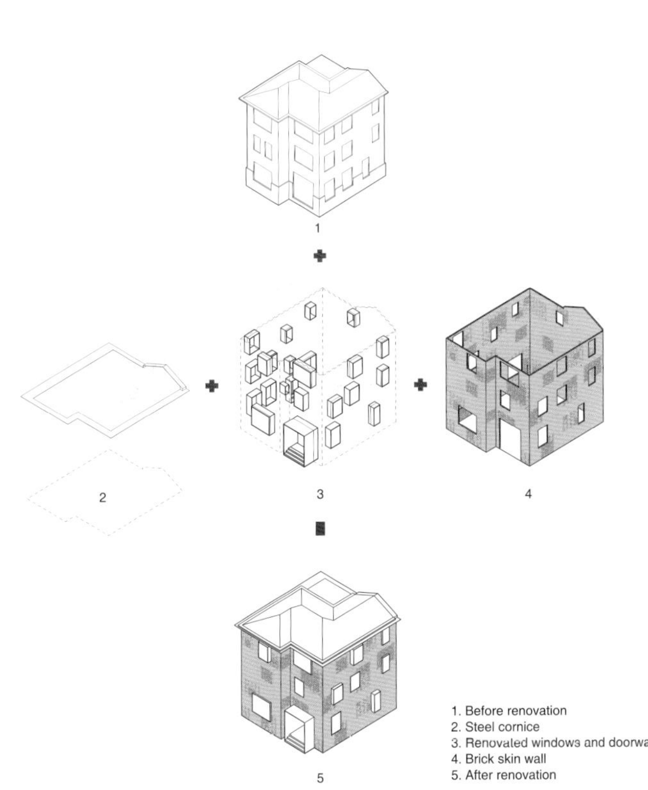

Diagram of facade renovation / 立面翻新示意图

Credits and Data

Project title: Facade Renovation for Building 8, Hengshanfang
Client: Shanghai Hengfu Properties, Co, Ltd.
Program: Commercial
Location: Xuhui District, Shanghai, China
Design: 2012
Completion: 2014
Status: Completed
Architects: Atelier Archmixing
Design Team: Zhuang Shen, Wang Kan, Yang Yunqiao, Xie Wenjing (Intern)
Illuminated brick manufacturer: Gainker (China) Building Technology Company
Structural mechanical and electrical engineer: Shanghai Sanyi Architectural Design company, Shanghai Yuangou Design and Consultant Company
Main contractor: Shanghai Xufang Construction Industry Company
Total floor area: 230 m²

Interview:

Building an Attitude for Chinese Architecture

Wang Shu
Interviewer: Li Xiangning

访谈：

建立一种中国建筑观

王澍
访谈者：李翔宁

All photos on pp. 156–159: Works by Wang Shu. This page, 2 photos: Xiangshan Campus, China Academy of Art, Hangzhou, China (phase I: 2004, phase II: 2007). Opposite, 2 photos: Ningbo History Museum, Ningbo, China (2008). p. 159, 2 photos: The Mountain residence by the waterside. All photos on pp. 156–159 courtesy of the architect.

156–159页的所有图：王澍的作品。本页，两图：杭州中国美术学院象山校区（一期：2004，二期：2007）。对页，两图：宁波历史博物馆（2008）。159页，两图：水岸山居。

Li Xiangning (LX): What has been your focus on architectural practices and thoughts since you won the Pritzker Architecture Prize in 2012? Would you please tell us about specific projects?

Wang Shu (WS): I am currently interested in lifestyles, particularly those in villages located between urban and rural areas. In China, there is no demand for detached house design in urban areas, but architects can engage in such designs for suburban areas. Considering that regional community formatic as well as house design is only possible in areas other than cities. I recently began the Wencun Village Project, and project in other suburban areas were also launched simultaneously. Unlike urban-style projects, these projects usually can only be initiated after building relationships and exchanging opinions with stakeholders in the area for at least one year. We wish to step forward to the next phase. In the previous phase the government will invest and contractors will proceed with construction based on our designs; while in the new phase, the design will still be arranged by us, but people at the site or the villagers are going to do the work by themselves. This phase is still being explored by everyone.

LX: Speaking of attitudes toward architecture, I once published an essay, "Architecture as Resistance". The same title was used for your solo exhibition in Belgium. When you discuss resistance, what is the purpose and subject? What degree of resistance can be translated into creation?

WS: I would rather avoid conscious discussion of these issues, but there is surely a thing to be called resistance in design. In particular, at the time when we live, people's survival, life, freedom, and other rightful conditions are under significant threats and are rapidly changing. I am resistant to this situation in other words, to industrial and commercial mega "systems" including behavioral manipulation and language operation. I resist through architecture, but there are other ways such as writings, paintings, etc. If I am to take a specific method, I would proceed to dissolve these mega systems by making every small effort and by a seemingly passive but actually quite secure method. That is the meaning of my existence.

LX: Today, the Chinese economy is said to be at a turning point. Do you think this will affect the architecture industry, including architectural culture?

WS: I think there would be no significant impact on us. The problems between the economy and the construction industry are a large amount of dead stocks in real estate, and the

翔宁：从获得普利兹克奖到现在的这几年，您的实践以及思考和关的问题是什么？有哪些具体在做的项目？

澍：我现在对人们的生活方式更感兴趣，这也是为什么我现在比较注乡村问题的原因。目前在中国，只有在乡村才会涉及诸如独栋住的这类设计，而在城市里实际上是没有的。不仅是住宅设计，社区形成在城市里也是几乎做不到的。我最近在做文村项目的同时，其的乡村项目也都在启动中。乡村项目不同于城市，它们一般都是与个地方先建立起联系，之后大概要经过一年甚至不止一年的来往过，项目才可能逐渐落地。接下来可能想尝试着将实践引入一个新阶。之前的项目是我们做设计，多由政府投资，之后有集中的公司进施工。而在新的阶段，项目的设计方案由我们提供，方案实施可能要靠当地人，甚至可能是村民自己来施工。这个阶段大家都还在摸。

李翔宁：说到对建筑的态度，我写过一篇文章叫《抵抗的建筑学》，来我注意到，您在比利时的一个展览中用这个作为您个展的题目。果说有一种抵抗的态度存在，那么抵抗的目的、对象是什么？抵抗何种程度上会成为一种创作的态度？

王澍：尽管我不想过于意识形态地去讨论这种问题，但设计中确有这种东西，尤其是在我们所处的这个各方面都在发生剧烈变化的时代。在这个时代里，人的生存、人的生活、人的自由，包括人的那种本应如其所是的存在受到了威胁。我的整个态度是针对着这种状态的，针对着那个巨大的生产和商业的系统，也包括行动操作的系统和话语操作的系统，即我们称之为“system”的这个东西，可以说它是我所有抵抗的对象。我可以用建筑的方式来做，可以换写作的方式、绘画的方式，也可以换其他行动的方式来做。用一种比较具象的方式表达，就是通过各种小努力、看起来非常消极但实际上非常坚定的方式，来对这个大系统做这种消解的工作。这是我存在的价值。

李翔宁：现在大家都说中国的经济到达了拐点，您认为这对建筑业有什么影响，会给建筑文化带来何种挑战？

王澍：对我们来说没有特别大的影响，因为这一轮在经济上与建筑有关的问题主要在于房地产大量积存，而城市的扩张在减速。这个对以

slowing speed of urban expansion. They might inflict heavy damage on the large-scale design institutes, whose clients are major developers, but they can be a chance for young architects and small-scale design offices. I remember that my first opportunity was also given during a recession. This is because mature business styles are favored during a booming economy, but everyone considers the possibility of economic activation through designs particularly during recessions. I have maintained a critical attitude toward the Chinese real estate industry, and have not been engaged in real estate design for more than ten years. The real estate industry is a business and is perhaps not contributing to cities in any way. For this matter, I think destruction of cities by the real estate industry is greater than its contribution, so I have avoided being involved with the industry for a long time.

LX: What do you think about the interpretation of tradition in modern Japanese culture? Today, many young architects are often inspired by Japanese architecture. Is there a difference or common understanding of each culture in China or in Japan?

WS: I think the two cultures have something in common. In particular, in the past, when architects intended to incorporate traditional elements in modern architecture in China, they referred to Japanese architecture, which began its attempts at integrating tradition earlier than in China and therefore had more experience. However, the circumstances have changed and today, such a situation has become a thing of the past. I think it is problematic that a culture is treated as a symbol for expressing modern architecture. If things become symbols or signs, a dialogue relationship between tradition and the modern can never be established. In a dialogue, a succession relationship should be developed in a type of continuity, where tradition is old and modern is new. Tradition and modern are different values and are issues with a different development pass. There are no differences as to which is modern and which is past. Recently, I had several opportunities to visit farming villages in China with international scholars. European scholars did not have the view of "this is Chinese tradition". What they saw was not tradition, but a feature. Hence, there was a misunderstanding in the discussion we've had so far. Today, we finally became able to recognize the current development and lifestyles in terms of values that are different from the past, i.e., traditions existed in China or Japan in the past. The relationships among traditions, nature, diversity, and initiative of life were dealt with well, but traditions are not rejecting large-scale and systematic developments. How to change direction by different values against "modernization", which occupies a major place in society? Or, if such an orbit adjustment is unlikely, realization of the coexistence of different values (whi[illegible] taking the present consumption-led values as a given) will lead to a possible opportunity to change direction.

Translated from Chinese by Haruki Mak[illegible]

Wang Shu, an architect and professor, was born in 1963 in Urumqi, a city in Xinjiang Uygur Autonomous Region, the western province of China. He receive[illegible] his Bachelor and Master's degree in architecture at the Nanjing Institute of Technology (now Southeast University). He received his Ph.D degree at Tongji University, 2000. He and his wife, Lu Wenyu, founded Amateur Architecture Studio in 1997 in Hangzhou, China. He is a Professor at China Academy of Art, Hangzhou, 2000; He became the Head of the Department of Architecture at China Academy of Art, 2003; and was named Dean of the School of Architectur[illegible] in 2007. In 2011, he became the first Chinese Kenzo Tange Visiting Professor a[illegible] the Harvard Graduate School of Design (GSD) in Cambridge, Massachusetts. He has been awarded a number of awards including Pritzker Architecture Prize in 2012 (the first Chinese citizen to win), the Gold Medal of Architecture (grande médaille d'or) from the l'Académie d'Architecture of France in 2011. He has participated in several international exhibitions including the 2006 and 2010 Venice Biennale of Architecture.

王澍，建筑师，教授。1963年出生于中国新疆维吾尔自治区的乌鲁木齐市。毕业于南京工学院（现在的东南大学），获建筑学学士与硕士学位。2000年毕业于同济大学，获博士学位。1997年与妻子陆文宇一起在杭州成立了"业余建筑工作室"，2003年成为中国美术学院建筑艺术学科带头人，2007年起担任建筑艺术学院院长。2011年成为首位在哈佛大学研究生院（GSD，位于美国马萨诸塞州的剑桥）担任"丹下健三客座教授"的中国本土建筑师。曾获得2012年的普利兹克建筑奖（第一位中国籍获奖人），2011年的法国建筑学院的建筑学金奖等一系列奖项。同时参与了一系列国际展览，包括2006年和2010年的威尼斯建筑双年展。

地产为主要工作领域的大型设计院无疑是一个比较大的冲击，而对些年轻的建筑师和小事务所，反而可能是很好的机会。我记得我自第一次开始有机会做项目的时候就是在经济不好的时候。经济好的候基本上是一些成熟的商业模式在推广，只有在经济不好的时候大才会觉得是不是可以通过设计来把某件事情激活。我已经有大概10多的时间不设计房地产项目了，因为我原则上对中国式的房地产持底的批评态度。我觉得那只是一种商业，它对城市几乎可以说是没任何贡献，甚至可能对城市的破坏远超对城市的贡献。所以，我从不去碰它。

翔宁：您怎么看待日本当代文化对于传统的解读？现在很多年轻的建筑师从日本建筑中获得了很多启示，您认为中国和日本文化对传统的认识有什么差异或共识？

澍：两边的文化确实是有渊源的。尤其是在新中国的早期，当建筑师想做现代建筑又想参考传统时，会比较多地借鉴日本，因为日本在这方面的探索比较早，他们有一些经验。但现在的情况已经不同了，那个阶段已经过去了。

我觉得真正的问题在于不应该把文化作为现在建筑表达当中的符号或者标志，因为这不是一个传统和现代的对话关系。那个对话里面隐含着“传统是过去的，现代是新的”这种潜意识，要建立起某种连续性的继承关系。今天我们认为那是不同的价值观的问题，是不同的发展路径的问题，它们之间不存在哪个现代哪个不现代的问题。我最近带了很多国际学者下乡，你会发现那些欧洲学者们不会有“这是中国的传统”这样的想法，他们看到的不是传统而是未来。所以，原来的那种讨论实际上是一种误读。我们今天可能会认识到，中国或日本也曾经有过的这样一种所谓的传统，它是另一种价值观上的发展和生活方式。它和自然、多样性、生活的自发性之间的关系被处理得更好，但同时它也不排除大规模的、有组织的发展。对“现代化”这样一个已经占社会主导地位的价值观，我们该如何用另一种价值观对其进行校正，或者说怎么让另外一种价值观至少在现在的消费主导的价值观的前提下，可与之共存，之后若有机会再对后者进行校正。

Chronicle of Chinese Architecture: 2004–2015

Compiled by Zhang Xiaochun, Ma Weidong, Wu Ruixiang

中国建筑年表：2004–2015年
汇编：张晓春 马卫东 吴瑞香

	Global Social Context	Major Events of Construction Industry in China	Major Projects
2004	• 3rd Anniversary of China's joining of the World Trade Organization (WTO). • February: Environmentalist NGO "Green Watershed" prevents the construction of the Nujiang hydroelectric project. • March: China introduces the concept of "human rights" into its Constitution for the first time. • September: The Communist Party of China sets the historic goal of "creating a harmonious society." • October: China proposes the creation of the East Asia Free Trade Area. • December: Tsunami takes place in the Indian Ocean.	• Ai Weiwei organizes exhibition and critique for the "Ordos 100" project (unrealized). • Chang Qing wins gold in the Asia Pacific region at the inaugurating Holcim Award for sustainable architecture. • Beginning of proposals for the Jianchuan Museum Complex in Sichuan, with three main themes – Cultural Revolution artworks, Second Sino-Japanese War (1937–1945), and Folklore & Culture – established. • The number of Class I certified architects in China reaches 12,664. • April: The former Ministry of Construction and Ministry of Science and Technology releases the reporting guideline for the national science and technology research program, launching a major national project for the 10th Five-Year Plan, "Research on Key Technologies for Green Architecture." • December: President Hu Jintao points out the need to develop energy and land saving housing, to fully promote and spread energy-saving technologies, and to set more stringent standards on saving water, energy, and materials.	• Shizilin Clubhouse, Beijing • Campus Complex, Donggu University of Technology • Yuhu Elementary School • Building C, College of Architecture and Urban Planning, Tongji University • Minsheng Bank Tower, Wuh • Abacus Museum, Nanjing
2005	• January: China's Kunlun Expedition for the Antarctic Inland Ice Sheet has confirmed to have found the highest point on the ice sheet. This was the first time that humans ever reached this highest point. • January: The European Parliament passed the European Constitution Treaty. • February: The Kyoto Accord, which limits global greenhouse gas emission, was finally agreed upon by 120 nations and put into effect after nearly eight years of dispute and debate. • March: The 3rd session of China's 10th National People's Congress was held, with Hu Jintao elected as the Chairman of the Central Military Commission of the State. • August: The first U.S.-China Strategic and Economic Dialogue takes place in Beijing. • October: China's Shenzhou 6 human spaceflight is successfully launched and landed, carrying a crew of two astronauts. • October: China proposes the development of a "new socialist countryside." • December: China and Japan sign the Sino-Japanese Long-term Trade Agreement in Tokyo.	• The inaugurating 1st Shenzhen-Hong Kong Bi-City Biennale of Urbanism / Architecture, themed *City, Open Door!* was opened. Curator: Yung Ho Chang. • Yuhu Elementary School designed by Chinese architect Li Xiaodong wins Gold Medal at the ARCASIA Awards for Architecture. • The 51st Venice Biennale was held in Venice, Italy. The Chinese Pavilion themed *Bamboo Shoot*, was curated by Yung Ho Chang. • Cheng Taining wins the 3rd Liang Sicheng Architecture Award, while Liu Keliang and Liu Li win nomination for the award. • The *a+u* magazine of Japan and the *El Croquis* magazine of Spain was introduced to publication in China by CA GROUP, and the *Architectural Record* of the U.S. was introduced to publication by China Architecture & Building Press. • Professor Wu Shuoxian of South China University of Technology becomes an academician of the Chinese Academy of Sciences. • Zhuang Weimin becomes Co-Chair at the UIA Professional Practice Commission. • Wang Xiaodong wins the Sir Robert Matthew Prize, International Union of architertional (UIA).	• Phase I of, Xiangshan Camp China Academy of Art • Memorial Hall of Victims in Nanjing Massacre by Japan Invaders, Renovation & Expansion Project • Anting New Town, Shanghai • World Trade Centre, Chongq • Olympic Sports Centre, Nanji • Desheng Noble Town, Beijing • Museum at Wangwushan-Daimeishan Global Geopark • Technology Tower at Tsinghu Science Park • Hebei Education Publishing House Tower
2006	• Full completion of the Three Gorges Dam. • The launch of a new round of real estate market regulation • March: The National People's Congress passes the plan abstract for the 11th 5-Year Plan. • April: Cross-Strait Economic, Trade and Culture Forum takes place in Beijing. • August: The super typhoon "Saomei" hits coastal areas in southeastern China. • September: Shinzo ABE becomes the Prime Minister of Japan. • October: BAN Ki-moon becomes the 8th Secretary-General of the United Nations. Shinzo ABE visits China, with the Chinese and Japanese leaders discussing further strategical and reciprocal relations.	• June-September: The China Contemporary architecture, art, and visual culture exhibition takes place in Rotterdam. • July: The Chinese edition of *DOMUS* begins publication. • September: the 10th Venice Architecture Biennale, with China joining in a national pavilion for the first time. Wang Shu: *Tiles Garden* • *MAD in China – a Practice About the Future*, opens as a collateral exhibition of the Venice Biennale. • Ma Qingyun becomes the dean of the School of Architecture at the University of Southern California. • Office Building of Qingpu Private Enterprise Association wins Commercial Building of the Year at the 2006 *Business Week / Architectural Record* China Awards and Honorable Mention at the WA Chinese Architecture Awards.	• Ningbo Museum • Zhujiajiao Government Centre, Qingpu District, Shanghai • Renovation of the Boyi Galle at the Ministry of Cultural Affairs, Beijing • Weifang Kite Square, Shandong • Jia Pingwa Literature Art Museum • Sino-French Centre, Tongji University • Vertical Courtyard Apartmer
2007	• March: The European Union celebrates its 50th Anniversary. • April: Wen Jiabao makes formal visits to Japan and Korea. Designated as the "China-Korea Exchange Year," the year marks the 15th anniversary of the establishment of formal diplomatic relations between China and Korea. • May: Cyanobacteria break out in the Wuxi-Taihu Lake region, which made tap water in nearly 1 million households undrinkable. • June: "China's National Policies and Actions for Addressing Climate Change" is released. • October: The crude oil price in international oil market reaches new heights, from which the world economy suffers. • October: The successful launch of China's first lunar-orbiting spacecraft, "Chang'e 1" • November: "Chang'e 1" sends back its first picture from the lunar orbit 380,000 km from Earth, marking the success of China's first lunar exploration project. • Known as the European Troika, Germany, France, and the U.K. all saw leadership changes, marking a new age of economic and social trends.	• Wang Xiaodong and Cui Kai win the 4th Liang Sicheng Architecture Award, while Chai Peiyi and Huang Xingyuan win nomination for the award. • The 2nd Shenzhen-Hong Kong Bi-City Biennale of Urbanism/Architecture (Shenzhen), themed *City of Expiration and Regeneration* opens. Curator: Ma Qingyun • Architect Cui Kai wins gold at ARCASIA's 2007–2008 Annual Architecture Awards, while Wang Ge wins Honorable Mention. • The selection process for design proposals for the China Pavilion at the 2010 Shanghai World Expo – organized by the Ministry of Construction, the Architectural Society of China, as well as the Shanghai Expo Group – finishes, narrowing down the final decision to three options. • The 2007 *Get It Louder* art exhibition held in Guangzhou, Shanghai, Beijing, and Chengdu for four months. • September: The Young Architects' Innovative Design National Summit Forum, themed *Innovative Design – New Architecture, New Hot Topics, New Concepts, New Technology*, takes place in Beijing. • December: The 2nd Holcim Forum for Sustainable Construction takes place in Tongji University.	• The new Suzhou Museum • National Centre for the Performing Arts • The new Capital Museum • Jinhua Architecture Park, Zhejiang • Jinchang Cultural Centre • The Health Centre of Nansha Grand Hotel
2008	• January: Most of southern China suffers from heavy ice and snow unseen in 100 years. • March: The opening of Shanghai's Land Transaction Market. • March: The 9th China Development Forum, themed "China 2020: Development Goals and Policy Approach," takes place in Beijing. • April: The Beijing-Shanghai High Speed Railway launches construction. • May: The Wenchuan Earthquake hits central China. • August: The 29th Summer Olympic Games takes place in Beijing. • September: The 13th Paralympics takes place in Beijing. • Successful launch of China's "Shenzhou 7" human spaceflight. • China's mainland and Taiwan realize the "Three Links," including postal, transportation, and trade links. • Barack Obama of the Democratic Party elected as the President of the U.S. • The subprime mortgage crisis and its ensuing financial crisis transmits globally and becomes a key financial and economic issue in the world.	• Chai Peiyi and Huang Xingyuan win the 5th Liang Sicheng Architecture Award, while Huang Xiqiu wins nomination for the award. • September: The 11th Venice Architecture Biennale was held in Venice, Italy. The Chinese Pavilion themed *Regular Architecture*, was curated by Yung Ho Chang. • April: The Architectural Society of China and the Human Resources and Education Department of the Ministry of Housing and Urban-Rural Development attend the 3rd International Architectural Education Accreditation/Validation Roundtable Conference and sign the Canberra Accord on Architectural Education – Recognition of Substantial Equivalence Between Accreditation/Validation Systems in Architectural Education. • April: *ARCHITopia 2*, an exhibition showcasing China's frontier architecture offices, opens in Brussels. • May: The Architectural Society of China holds experts' panel after the Wenchuan Earthquake. • June: The National Council of the Order of French Architects (CNOA) hosts an exhibition in Paris for Chinese contemporary architecture, themed *Location*. • October: The 7th International Symposium on Architectural Interchanges in Asia takes place in Beijing. Theme: "Urban Regeneration and Architecture Creation". • October: The 3rd China International Architectural Biennale takes place in Beijing. • November: The *Feng Jizhong and Fangta Garden* exhibition and academic symposium take place in Shenzhen. • December: Winners for the 1st China Architecture Media Awards released, with Maosi Elementary School wins the Best Architecture Award. • December: Editorial departments of the *Architectural Journal, Journal of Building Structures*, and *a+a* Magazine were converted from official agencies to magazine corporations.	• National Stadium, Beijing (Bird's Nest) • National Aquatics Centre, Beijing (Water Cube) • Control Centre for the 2008 Olympic Games • Phase II of Xiangshan Campus China Academy of Art • Urban Tulou • Yaluntzangpu Boat Terminal • Phase II of the National Library of China • Terminal 3 of Beijing Capital International Airport • Phoenix Media Tower, Shenzhen • Concrete Slit House • Protective restoration project c the "Sangzhu Zizong Castle" ir Shigatse, Tibet
2009	• The Cultural Centre at CCTV (China Central Television)'s new headquarters catches fire on the night of the Lantern Festival. • February: Premier Wen Jiabao communicates with netizens online. This is the first time that the Chinese government's premier communicates with the public online in real-time. • July: The "7·5 Rioting Serious Violent Criminal Incident" takes place in Urumqi. • April: The Influenza A H1N1 spreads across the world. • April: New plans for the healthcare system reform released. • December: The 2009 United Nations Climate Change Conference held in Copenhagen. • The world economy reestablishes new order after the financial crisis.	• Yung Ho Chang becomes the Head of the Department of Architecture at MIT. • Chang Qing named the Honorary fellowship, the American Institute of Architecture (Hon. FAIA). • Neri & Hu Design and Research Office wins Design for Asia Awards (DFAA). • The 3rd Shenzhen-Hong Kong Bi-City Biennale of Urbanism /Architecture (Shenzhen) takes place. Theme: *City Mobilization*. Curator: Ou Ning. • World Design Congress 2009 Beijing as well as the inaugurating Beijing Design Week take place, with main themes *Design·Innovation·Economy*. • May: Award Ceremony for Architectural Retrospect and Creation in 60 Years of Chinese History takes place in Shanghai. • June: The 2009 China Sustainable Architecture Conference takes place in Shanghai. • October: The 1st China (Haixi) Eco-Habitat Summit Forum takes place in Xiamen.	• Linked Hybrid (Dangdai MOMA), Beijing • Shanghai World Financial Centre • Ningbo History Museum • Hutong Bubble 32 • Xi'an Television & Broadcasting Centre • The Water House, Lijiang; School Bridge • Urban Planning Exhibition Hall of Jiading New Town, Shanghai • Vanke Centre, Shenzhen

世界 / 中国大事件	中国建筑界的主要事件	主要建筑作品
国加入世界贸易组织(WTO) 三周年。 月，NGO环保组织阻建怒江水电站。 月，中国首次将"人权"概念引入宪法。 月，中国共产党正式提出了"建立和谐社会"的历史目标。 0月，中国提出建立东亚自由贸易区。 2月，印度洋发生海啸。	• 艾未未组织"鄂尔多斯100"项目展评(未实现)。 • 常青(作品：杭州来氏聚落)获得首届赫尔希姆(Holcim)国际可持续建筑大奖赛亚太区金奖。 • 四川"建川博物馆群"开始提交方案，博物馆规划设计了"文革"艺术品博物馆、抗战博物馆和民间博物馆三条主线。 • 中国一级注册建筑师达到12,664人。 • 4月，原建设部与科技部发布了国家科技攻关计划重点项目申报指南，启动了"十五"国家科技重大攻关项目——"绿色建筑关键技术研究"。 • 12月，胡锦涛指出，要大力发展节能省地型住宅，全面推广和普及节能技术，制定并强制推行更严格的节能节材节水标准。	• 北京柿子林会所 • 东莞理工大学建筑群 • 玉湖完小 • 同济大学建筑与城市规划学院C楼 • 武汉民生银行大厦 • 南通珠算博物馆
月，中国南极内陆冰盖昆仑科考队确认找到南极内陆冰盖的最高点，是人类首次登上南极内陆冰盖最高点。 月，欧洲议会通过欧盟宪法条约。 月，限制全球温室气体排放量的《京都议定书》经过近8年争拗后，终获得120多个国家确认，并正式生效。 月，十届全国人大三次会议召开，选举胡锦涛为国家中央军委主席。 月，中美首次战略对话在北京举行。 0月，中国载有两名航天员的神舟六号载人飞船成功发射并顺利着陆。 0月，中国提出"社会主义新农村建设"。 2月，中国和日本在东京签署中日长期贸易协议。	• 首届"深港城市\建筑双城双年展"开幕，主题：城市，开门！ 策展人：张永和。 • 中国建筑师李晓东设计的"玉湖完小"获得亚洲建筑金奖。 • 意大利威尼斯艺术双年展举办，张永和策展"竹跳"。 • 程泰宁获第3届梁思成建筑奖，刘克良和刘力获第3届梁思成建筑奖提名奖。 • 日本的《建筑与都市》(A+U)和西班牙的《建筑素描》(EL CROQUIS)由文筑国际、美国的《建筑实录》(Architectural Record)由中国建筑工业出版社引进中国出版。 • 华南理工大学建筑学院吴硕贤教授当选为中国科学院院士。 • 庄惟敏任国际建筑协会建筑职业实践委员会联席主任。 • 王晓东获国际建协改善人类居住质量专业奖。	• 中国美院象山校区一期 • 侵华日军南京大屠杀遇难同胞纪念馆改扩建工程 • 上海安亭新镇 • 重庆世界贸易中心 • 南京奥体中心 • 北京德胜尚城 • 王屋山世界地质公园博物馆 • 清华科技园科技大厦 • 河北教育出版社大厦
三峡大坝全面建成。 新一轮房地产调控启动。 月，全国人民代表大会批准"十一五"规划纲要 。 月，两岸经贸论坛在京举行。 月，超强台风"桑美"重创中国东南沿海地区。 月，安倍晋三当选日本首相。 10月，潘基文就任联合国第8任秘书长。安倍晋三访华，中日两国领导人就构筑中日战略互惠关系达成一致。	• 6月–9月，鹿特丹举办"当代中国'建筑、艺术与视觉文化大展"。 • 7月，意大利DOMUS杂志国际中文版出版。 • 9月，第10届威尼斯建筑双年展，中国首次以国家馆形式参展，王澍：瓦园。 • MAD in China——一个关于未来的实践，威尼斯建筑双年展外围展。 • 马清运担任美国南加州大学建筑规划学院院长。 • "青浦私营企业协会办公与接待中心"获2006 年美国《商业周刊》/《建筑实录》评选的最佳商用建筑奖和2006 WA 中国建筑奖佳作奖。	• 宁波美术馆 • 上海青浦朱家角行政中心 • 北京文化部博艺画廊改建项目 • 山东潍坊风筝广场 • 贾平凹文学艺术馆 • 同济大学中法中心 • 垂直院宅
3月，欧盟庆祝成立50周年。 4月，温家宝对韩国、日本进行正式访问。该年是中韩建交15周年，也是中韩交流年。 5月，太湖无锡流域大面积蓝藻暴发，近百万市民家中的自来水无法饮用。 6月，《中国应对气候变化国家方案》正式颁布。 10月，国际原油价格创新高，世界经济遭受"高油价之痛"。 10月，中国第一颗探月卫星"嫦娥一号"发射成功。 11月，"嫦娥一号"从距离地球38万km的环月轨道传回第一张月面图片，标志着中国首次探月工程取得圆满成功。 号称欧洲"三驾马车"的德、法、英三国相继出现新领军人，欧洲进入新"三剑客"时代。	• 王晓东、崔愷获第4届梁思成建筑奖，柴裴义、黄星元获第4届梁思成建筑奖提名奖。 • 第2届"深港城市\建筑双城双年展(深圳)"举办，主题：城市再生，策展人：马清运。 • 崔愷获亚洲建协金奖(2007–2008年度建筑奖)，王戈获亚洲建协荣誉奖。 • 由建设部、中国建筑学会和上海世博集团共同主持的"中国2010年上海世博会中国馆项目建筑方案"评审工作结束，确定3个方案为推荐方案。 • "2007大声展"在广州、上海、北京，成都举办，历时4个月。 • 9月，以"创新设计——新建筑、新热点、新理念、新技术"为主题的全国青年建筑师创新设计高峰论坛在北京举行。 • 12月，第2届Holcim全球可持续建筑国际论坛于同济大学举办。	• 苏州博物馆新馆 • 国家大剧院 • 首都博物馆新馆 • 浙江金华建筑艺术公园 • 金昌市文化中心 • 南沙大酒店健康中心
1月，中国南方大部分省份遭遇了百年难遇的冰雪灾害。 3月，上海市土地交易市场正式开业。 3月，第9届中国发展高层论坛在北京召开。中国发展高层论坛的主题是"中国2020：发展目标和政策取向"。 4月，京沪高速铁路全线开工。 5月，四川省汶川市发生大地震。 8月，第29届夏季奥林匹克运动会在北京举办。 9月，第13届残奥会在北京举办。 中国"神舟七号载人航天飞船"成功发射。 中国海峡两岸基本实现三通。 美国民主党奥巴马当选美国总统。 美国次贷问题以及引发的金融风暴向世界各国传导，成为世界金融和经济问题。	• 柴裴义和黄星元获第5届梁思成建筑奖，黄锡璆获第5届梁思成建筑奖提名奖。 • 9月，"威尼斯建筑双年展"举办，中国馆主题：普通建筑，策展人：张永和。 • 4月，中国建筑学会与住房和城乡建设部人事教育司参加了第3次国际建筑教育评估认证圆桌会议，并签署了《建筑学教育评估认证体系间实质对等性承认协议》。 • 4月，布鲁塞尔举行"建筑乌托邦2"中国新锐建筑事务所设计展。 • 5月，汶川大地震后，中国建筑学会组织建筑专家会议。 • 6月，法国建筑师学会在巴黎举办主题为"位置"的当代中国建筑展。 • 10月，第7届亚洲建筑国际交流会在北京举办，主题为"城市更新与建筑创新"。 • 10月，第3届中国国际建筑展在北京举办。 • 11月，"冯纪忠和方塔园"展览暨学术研讨会于深圳举行。 • 12月，首届中国建筑传媒奖获奖名单揭晓："毛寺生态实验小学"获最佳建筑奖。 • 12月，《建筑学报》、《建筑结构学报》和《建筑知识》的3个编辑部先后完成"杂志社有限公司"的体制改革。	• 鸟巢国家体育场 • 水立方国家游泳中心 • 2008奥运会控制中心 • 中国美术学院象山校区二期 • 土楼公舍 • 雅鲁藏布江小码头 • 国家图书馆二期 • 北京首都国际机场T3 航站楼 • 中国凤凰传媒大厦 • 混凝土缝之宅 • 西藏日喀则"桑珠孜宗堡"保护性修复工程
元宵节，在建央视新办公楼的文化中心发生大火。 2月，温家宝与网民在线交流，这是中国政府总理首次与网民进行实时交流。 4月，甲型H1N1流感肆虐全球。 4月，新医改方案出台。 7月，乌鲁木齐发生"7·5"打砸抢烧严重暴力犯罪事件。 12月，哥本哈根气候大会举行。 经济危机后，世界经济秩序重建。	• 张永和就任美国麻省理工学院(MIT)建筑系主任。 • 常青被选为美国建筑师学会荣誉会士(Hon.FAIA)。 • 如恩设计获得亚洲最具影响力设计大奖(DFA)。 • 第3届"深港城市\建筑双城双年展(深圳)"举办，主题：城市动员，策展人：欧宁。 • 2009北京世界设计大会暨首届北京国际设计周举办，活动主旨为"设计·创新·经济"。 • 5月，"新中国60年建筑回顾和建筑创作大奖颁奖典礼"在上海举行。 • 6月，"2009中国可持续建筑国际大会"在上海举办。 • 10月，"首届中国(海西)生态人居高层论坛"在厦门举办。	• 北京当代MOMA • 上海环球金融中心 • 宁波博物馆 • 胡同泡泡32号 • 西安广播电视电影中心 • 淼庐、桥上书屋 • 上海嘉定新城规划展示馆 • 深圳万科中心

	Global Social Context	Major Events of Construction Industry in China	Major Projects
2010	• The 2010 World Expo successfully held in Shanghai. • January: Earthquake hits Haiti, marking its most serious disaster in 200 years. • April: The Yushu Earthquake hits Qinghai Province. • April: The Gulf of Mexico Oil Spill takes place in the Mississippi River Delta in the United States. • August: The Zhouqu Mudslide hits Gansu Province. • July: Pakistan suffers from serious flood, covering almost the entire Indian River basin. • Summer: The FIFA World Cup held in South Africa. • November: The 16th Asian Games held in Guangzhou. • November 15th: An apartment on Shanghai's Jiaozhou Road was caught on fire, striking conversations and considerations on building safety concerns such as disaster prevention, insulation, and fireproofing.	• The Bridge School by Li Xiaodong wins the Aga Khan Award for Architecture. • "Three Narratives About the Past Ten Years" creation and retrospect exhibition takes place in Beijing. Sponsor: *The Architect* magazine, China. • Launch of the 2nd "China Architecture Media Awards." • UIA Zone Four's Architectural Heritage Preservation International Conference takes place in Xi'an. • The 2010 Annual Conference of the Architectural Society of China, themed "Expo Architecture: Green and Innovation Design," takes place in Shanghai. • August: The 4th National Architectural Design Innovation Summit and the 2nd Shandong Province Green Architecture Summit take place in Yantai. • September: The Architectural Society of China organizes a team of best architects and design teams across the country in support of design and planning works for major post-quake construction projects in Yushu. • October: The 5th China International Architectural Biennale and the 2010 China Architectural Forum take place in Beijing. • November: The 1st International Forum on Chinese Ecological Habitat takes place in Hainan, passing the Ding'an Manifesto of the 1st (2010) International Forum on Chinese Ecological Habitat. Theme: Low-Carbon Ecological Housing and Archi.	• Shanghai Expo Complex: China Pavilion, Theme Pav Spain Pavilion, UK Pavilion etc. • Museum of Handcraft Pape Yunnan • Guangzhou Opera House • The new Guangdong Muse • The Waterhouse at South B • Giant Interactive Group Corporate Headquarters
2011	• 2011 marks the opening of China's 12th Five-Year Plan. • March: The Tohoku Earthquake and Tsunami hit Japan, resulting in the Fukushima Nuclear Power Plant crisis. • July: Two high-speed trains collide and derail on a viaduct in the suburbs of Wenzhou, Zhejiang. • September: The State Council releases "China Aging Development Plan for the 12th 'Five-Year Plan,'" as well as "Special Architectural Plan for Senior Care Facilities, 12th Five-Year Plan (2011–2015)". • The Fire Department at the Ministry of Public Security releases a document that further clarifies the regulations and requirements over the use of insulation and fireproof materials on building facades. • Continued regulation of the real estate market; construction begins for over 10 million units of social security housing. • 90th Anniversary of the founding of the Communist Party of China. • Commemoration Ceremony for the 100th Anniversary of the 1911 Revolution takes place. • The North Korean leader Kim Jung-il passes away, creating uncertainties for relations between North and South Korea. • The world population exceeds 7 billion. • The European debt crisis continues to impact the region, resulting in political changes in certain countries.	• The Builders' Apartments of the Sino-Singapore Tianjin Eco-City wins nomination for the UIA Vassilis Sgoutas Prize. • Cui Kai and Liu Jiaping become academicians of the Chinese Academy of Engineering. • The 4th Shenzhen-Hong Kong Bi-City Biennale of Urbanism/Architecture (Shenzhen) takes place. Theme: *Architecture creates cities. Cities create architecture.* Curator: Terence Riley • The 2011 Chengdu Biennial International Architecture Exhibition, themed *Changing Vistas, Creative Duration: Countryside / City / Architecture*, Chief Curator: Zhi Wenjun. • International Bidding Competition for the new National Art Museum of China. • "Bishan Project," aiming at bringing art to the countryside, launches at Times Museum in Guangzhou. Curator: Ou Ning • Yung Ho Chang becomes the first Chinese jury of the Pritzker Architecture Prize. • Atelier Deshaus was recognized by *Architectural Record* as one of the 10 "Design Vanguard 2011" around the world. • The exhibition "People's Architecture: Hsieh Ying-chun, Architect" takes place in Beijing, Shenzhen, Hong Kong, and Shanghai. • The 2011 Beijing Design Week and the 1st Beijing International Design Triennial take place in Beijing. Theme: *Design Beijing.*	• Renovation and expansion the National Museum of Ch • Ordos Museum • Yarlung Zangbo Grand Can Art Centre, Tibet • Oriental Sports Centre, Shanghai • Xi'an International Horticultu Exhibition Complex • LiYuan Library • Enjoy Museum of Art • Tangshan Museum • Huludao Beach Exhibit Cent
2012	• January: The National Bureau of Statistics' release shows that the number of urban population in mainland China (excluding Hong Kong, Macau, and Taiwan) exceeds that of the rural population, reaching 51.27%. • July: Beijing suffers from serious heavy rain and landslide disasters. • August: Russia officially becomes the 156th member state of the World Trade Organization (WTO). • October: Chinese author Mo Yan wins the Nobel Literature Prize. • November: The 18th National Congress of the Communist Party of China takes places, electing Xi Jinping as the General Secretary. • December: The Central Economic Work Conference proposes visions for a steady development of urbanization with high quality. • The 30th Summer Olympic Games held in London.	• Wu Liangyong wins the 2011 State Preeminent Science and Technology Award of China. • Wang Shu becomes the first Chinese winning the Pritzker Architecture Prize, with the Award Ceremony held in Beijing. • Chinese Pavilion at the Venice Biennale. Theme: *Originaire*. Curator: Fang Zhenning • October: Liu Li and Huang Xiqiu win the 4th Liang Sicheng Architecture Award, while Meng Jianmin and Tao Zhi win nomination for the award. • From research to design: the Chinese architects exhibition at the Milan Triennale. Curator: Li Xiangning • The 2012 Beijing Design Week takes place. Theme: *Upgrading the city by design.* • Jean Nouvel wins bid for the new National Museum of Art.	• The new CCTV (China Centra Television) headquarters rebu after fire • Phoenix International Media Centre, Beijing • Foshou Lake Architectural Complex, Nanjing • CIPEA No. 4 House • OCT Design Museum • White-Walled House – Imperia Street of Southern Song Dynasty renovation project • OCT Loft Creative Culture Par
2013	• July: Premier Li Keqiang proposes a new type of urbanization that centres around people. • September: China (Shanghai) Free-Trade Zone established. • September: The State Council releases "Atmospheric Pollution Prevention Action Plan," marking the beginning of a battle with air quality. • November: Typhoon "Haiyan" hits the Philippines. • December: Nelson Mandela, South Africa's former president, passes away. • December: "Chang'e 3" makes its first successful soft landing and investigation on the moon, marking new milestone for China's lunar exploration program.	• The 2013 West Bund Architecture and Contemporary Art Biennale takes place. Theme: *Reflecta and Fabrica*. General curator: Yung Ho Chang. Architecture curator: Li Xiangning. Art curator: Gao Shiming. • The 5th Shenzhen-Hong Kong Bi-City Biennale of Urbanism/Architecture (Shenzhen) takes place. Theme: *Urban Border*. Curators: Li Xiangning, Jeffrey Johnson, Ole Bouman • The National Maritime Museum of China wins Future Projects Award at the 2013 World Architectural Festival (WAF). • The Stone Buildings at the Northeast Dianli University designed by the Liang Sicheng couple becomes a nationally protected heritage site. • The new CCTV (China Central Television) headquarters named "Best Tall Building Worldwide" by the Council on Tall Buildings and Urban Habitat (CTBUH). • The 2013 Beijing Design Week takes place. Theme: *Capital of Design · Smart City.* • May: The 9th China International Garden Expo opens in Beijing. • November: The "2013 China Contemporary Architecture Design and Development Strategic International Forum" takes place at Southeast University in Nanjing.	• Wenchuan Earthquake Epicentre Memorial • Jianamani Visitor Centre • Jiading Public Library and Culture Centre, Shanghai • China Fortune Exhibition Cen • Micro-Hutong, Beijing • Jade Museum • Silo-top Studio • Niang'ou Boat Terminal, Tibet
2014	• Xi Jinping unveils the "Belt and Road" (Silk Road Economic Belt and the 21st-century Maritime Silk Road) development strategy, with the National Development and Reform Commission outlining the general proposal. • February: Xi Jinping elevates the importance of a joint development between Beijing-Tianjin-Hebei to a national level. • February: The Ebola epidemic breaks out in Guinea, the global community assists in controlling its impact. • March: Flight MH370 of Malaysia Airlines from Kuala Lumpur to Beijing goes missing. • November: The 26th annual gathering of APEC leaders takes place in Beijing. • November: China and Japan reach agreement in four principal points in improving relations between two nations. • December: The 1st phase of the central route of the North-South Water Transfer Project opens.	• Launch of the new National Museum of Art of China. • Ma Yansong wins design for the Lucas Museum of Narrative Art – this is the first time a Chinese architect wins design competition for a cultural landmark abroad. • The Long Museum wins second place of 2014 Architectural Review Emerging Architecture awards of *Architectural Record* and silver for the Design for Asia Awards (DFAA). • LiYuan Library designed by Li Xiaodong wins the inaugural Moriyama RAIC International Prize. • The 2014 Beijing Design Week takes place. Themes: *Capital of Design · Smart City · Ecology and Culture* • The Ministry of Housing and Urban-Rural Development releases notice for the Code for Fire Protection Design of Buildings • February: Xi Jinping inspects Beijing's urban planning, proposing that mistakes made in planning results in most serious waste. • February: The Shanghai Xiandai Architectural Design Group acquires Wilson Associates, an interior architectural design firm in U.S.. • March: The National New-Type Urbanization Plan (2014–2020) released. • August: Opening of the Shihlien Chemical Plant office buildings – Alvaro Siza's first project in China. • October: Xi Jinping proposes "no more weird architecture". • November: The State Council cancels the accreditation for Class I registered architects.	• Long Museum West Bund • Shandong Art Museum • Jixi Museum • Folk Art Museum, China Academy of Art, Hangzhou • Shanghai Poly Grand Theatre • Kunshan Eco-Farm Visitor Centre • Beijing No. 4 High School Fangshan Campus • Sun Farming Commune, Hangzhou • Z Gallery at iD Town • Wuyishan Bamboo Raft Factory • The Building on the Water, Shihlien Chemical
2015	• Premier Li Keqiang points out in government report that the construction of the new countryside should benefit all farmers, and that the construction of infrastructure shall be strengthened, creating a countryside that is beautiful and livable. • 70th anniversary of the victory of the World Anti-Fascist War. • September: The 2015 China Victory Day Parade (commemorating 70th anniversary of the victories of the World Anti-Fascist War) takes place in Beijing. • October: The Chinese pharmaceutical chemist Tu Youyou wins the 2015 Nobel Prize in Physiology or Medicine. • November: Ten countries of the Association of Southeast Asian Nations (ASEAN) have allied together to establish the ASEAN Economic Community (AEC). • November: Terrorist attack takes place in Paris. • December: The World Internet Conference takes place in Wuzhen. • December: The 21st United Nations Conference on Climate Change takes place in Paris, with the signing of the Paris Agreement.	• Chang Qing becomes an academician of the Chinese Academy of Sciences; Wang Jianguo and Meng Jianmin become academians of the Chinese Academy of Engineering. • Wu Jiang becomes a member of the Academie d'Architecture of France. • Chang Qing's protective restoration project of the "Sangzhu Zizong Castle" in Shigatse, Tibet wins Gold Medal at the ARCASIA Awards for Architecture. • The 2015 Beijing Design Week takes place. Themes: *Design Capital · Smart City · Industry Fusion.* • Cail Yongjie's Beichuan Earthquake Memorial wins Gold Medal at the ARCASIA Awards for Architecture. • The *Piece by Piece* exhibition by Renzo Piano Building Workshop opens in Shanghai. • September: Opening of the first Shanghai Urban Space Art Season. Theme: *Urban Regeneration.* Chief Curators: Wu Jiang, Mohsen Mostafavi. Architectural Curator: Li Xiangning, Art Curator: Zhang Qing. • December: The 6th Shenzhen-Hong Kong Bi-City Biennale of Urbanism/Architecture (Shenzhen) takes place in Shenzhen. Theme: *Re-living the city.* Curators: Aaron Betsky, Alfredo Brillembourg, Hubert Klumpner, Doreen Heng Liu. • Mei Hongyuan named the Honorary Fellow, College Fellows, the American Institute of Architecture (Hon. FAIA).	• China Pavilion at the Milan Expo • Shanghai Tower • Seashore Library • Harbin Opera House • Shanghai Natural History Museum • Minsheng Museum of Modern Art, Beijing

Translated from Chinese by George Huaiyu Zhang

世界 / 中国大事件	中国建筑界的主要事件	主要建筑作品
海世界博览会成功举行。 ，海地地震，为海地200年来最严重灾害。 ，青海省玉树市发生地震。 ，美墨西哥湾漏油酿生态灾难。 ，巴基斯坦洪灾，淹没了几乎整个印度河流域。 ，甘肃舟曲遭特大泥石流灾害。 季，南非世界杯举行。 月，第16届亚运会在广州举行。 月15日，上海胶州路建筑大火，引发建筑防灾、建筑保温材料安全等问题的思考。	• 李晓东设计的“桥上书屋”获得阿卡汗建筑奖。 • “东西南北中——十年的三个民间叙事”创作回顾展于北京举办，主办方：《建筑师》杂志。 • 第2届“中国建筑传媒奖”启动。 • “UIA亚奥地区(第四区)建筑遗产保护国际会议”在西安举行。 • 5月，以“世博建筑——绿色创新设计”为主题的2010中国建筑学会学术年会在上海举办。 • 8月，“第4届全国建筑设计创新高峰论坛暨第4届山东省绿色建筑设计高峰论坛”在山东烟台举办。 • 9月，中国建筑学会组织全国最强的设计团队和著名设计大师分别对玉树重点工程项目进行规划和设计工作。 • 10月，“第5届中国国际建筑展暨2010中国建筑论坛”在北京举办。 • 11月，以“低碳环保住房与建筑”为主题的“第1届中国生态人居国际论坛”在海南举办，会议通过了《第1届(2010)中国生态人居国际论坛定安宣言》。	• 上海世博会建筑群：中国馆、主题馆、西班牙馆、英国馆等 • 高黎贡手工造纸博物馆 • 广州歌剧院 • 广东省博物馆新馆 • 水舍精品酒店 • 巨人集团上海联合总部
011年是中国“十二五”计划的开端年。 月，日本9级地震引发海啸，同时造成福岛核电站危机。 月，温州鹿城区发生两动车追尾事故。 月，国务院发布《中国老龄事业发展“十二五”规划》和《“十二五”社会老服务体系建筑规划(2011-2015)》。 安部消防局发出《关于进一步明确民用建筑外保温材料消防监督管有关要求的通知》。 持房地产调控，千万保障房开始建设。 国共产党成立90周年。 亥革命100周年纪念大会举办。 鲜最高领导金正日去世，朝韩局势不明。 界人口突破70亿大关。 债危机持续蔓延，引发部分欧元区国家政坛更迭。	• “天津中新生态城建设者公寓”获国际建协斯古塔斯奖提名奖。 • 崔愷和刘加平当选中国工程院院士。 • 第4届“深港城市\建筑双城双年展(深圳)”举办，主题：城市创造，策展人：Terence Riley。 • “2011成都双年展国际建筑展”举办，主题为“物我之境：田园/城市/建筑”，主策展人：支文军。 • 中国国家美术馆新馆举行国际竞标。 • 艺术下乡项目“碧山计划”在广州时代美术馆正式启动，策展人：欧宁。 • 张永和成为普利兹克奖评委，是该奖项的首位中国评委。 • 大舍建筑设计事务所被美国建筑师协会会刊《建筑实录》评选为年度全球10佳“设计先锋”(Design Vanguard 2011)。 • “人民的建筑·谢英俊建筑师巡回展”在北京、深圳、香港和上海举行。 • “2011北京国际设计周暨首届北京国际设计三年展”举办，主题为“设计北京”。	• 中国国家博物馆改扩建项目 • 鄂尔多斯博物馆 • 西藏雅鲁藏布大峡谷艺术馆 • 上海东方体育中心 • 西安世界园艺博览会建筑群 • 篱苑书屋 • 悦美术馆 • 唐山博物馆 • 葫芦岛海滨展示中心设计
月，中国国家统计局公布数字，大陆城镇人口(不包括港澳台地区)超过农村人口，城镇人口占总人口比重达到51.27%。 月，北京遭遇特大暴雨山洪泥石流灾害。 月，俄罗斯正式成为世界贸易组织第156个成员国。 0月，中国作家莫言获诺贝尔文学奖。 1月，中国共产党第十八次全国代表大会召开，选举习近平为中央委员会总书记。 2月，中央经济工作会议提出，2013年要积极稳妥地推进城镇化，提高城镇化质量。 第30届夏季奥林匹克运动会在伦敦举行。	• 吴良镛获得2011年度国家最高科学技术大奖。 • 王澍获2012年普利兹克建筑奖，成为首获该奖的中国人，颁奖仪式在北京举行。 • 威尼斯建筑艺术双年展中国馆：“原初”展，策展人：方振宁。 • 10月，刘力和黄锡璆获第6届梁思成建筑奖，孟建民、陶郅和唐玉恩获第6届梁思成建筑奖提名奖。 • “米兰三年展：中国建筑师展”举办，策展人：李翔宁。 • “2012年北京国际设计周”启动，主题为“设计提升城市品质”。 • 让·努维尔中标中国国家美术馆新馆项目。	• CCTV新大楼火灾后重建项目 • 北京凤凰国际传媒中心 • 南京佛手湖建筑群 • CIPEA 4# 住宅 • OCT设计博物馆 • 白色墙屋-南宋御街改建项目 • OCT Loft 华侨城创意文化园
7月，李克强提出“推进以人为核心的新型城镇化”。 9月，中国(上海)自由贸易试验区正式成立。 9月，国务院发布《大气污染防治行动计划》。 11月，台风“海燕”重创菲律宾。 12月，南非前总统曼德拉逝世。 12月，“嫦娥三号”首次实现月球软着陆和月面巡视勘察，开启探月工程新的征程。	• “西岸2013建筑与当代艺术双年展”举办，主题：Reflecta (进程)和Fabrica (营造)，总策展人：张永和。建筑展策展人：李翔宁。艺术展策展人：高士明。 • “第5届深港城市\建筑双城双年展(深圳)”举办，主题：城市边缘，策展人：李翔宁、Jeffrey Johnson、Ole Bouman。 • “中国国家海事博物馆”获2013年世界建筑节(WAF)未来奖。 • 梁思成夫妇设计的“石头楼”被列入国家保护单位。 • “CCTV大楼”获2013年度全球最佳高层建筑奖。 • “2013北京国际设计周”启动，主题为“设计之都·智慧城市”。 • 5月，“第9届中国国际园林博览会”在北京开幕。 • 11月，“2013中国当代建筑设计发展战略国际高端论坛”在南京东南大学举办。	• 汶川大地震震中纪念馆 • 嘉那嘛呢游客到访中心 • 上海嘉定图书馆文化馆 • 华鑫展示中心 • 北京微胡同 • 卜石艺术馆 • 麦仓工作室 • 西藏娘欧码头
习近平提出“一带一路”发展战略，国家发改委制定“一带一路”总规划。 2月，习近平指出京津冀协同发展上升为重大国家战略。 2月，几内亚暴发埃博拉疫情，全球协力防控埃博拉疫情。 3月，从吉隆坡飞往北京的马来西亚航空公司“MH 370”航班失踪。 11月，亚太经合组织(APEC)第22次领导人非正式会议在北京举行。 11月，中日就处理和改善两国关系达成四点原则共识。 12月，南水北调中线一期工程正式通水。	• 中国国家美术馆新馆建筑设计项目正式启动。 • 马岩松获得卢卡斯叙事艺术博物馆设计权，这是中国建筑师首获海外文化地标设计权。 • 龙美术馆西岸馆获2014AR Awards for Emerging Architecture的二等奖以及年度亚洲最具影响力设计大奖(DFAA)环境设计类文化空间银奖。 • 李晓东设计的“篱苑书屋”获MoriyamaRAIC国际奖。 • 2014北京国际设计周举行，主题为“设计之都·智慧城市·生态文明”。 • 住建部发布国标《建筑设计防火规范》公告。 • 2月，习近平考察北京规划，提出“规划失误是最大的浪费”。 • 2月，上海现代建筑设计集团收购美国威尔逊室内设计公司。 • 3月，《国家新型城镇化规划(2014-2020)》正式对外发布。 • 8月，Álvaro Siza在中国的首个作品——实联化工办公楼正式启用。 • 10月，习近平提出“不要再搞奇奇怪怪的建筑”。 • 11月，国务院取消一级注册建筑师执业资格认定。	• 龙美术馆西岸馆 • 山东美术馆 • 绩溪博物馆 • 中国美术学院民艺馆 • 上海保利大剧院 • 昆山有机农场游客互动中心、采摘亭 • 北京四中房山校区 • 临安太阳公社竹构系列 • iD Town 之折艺廊 • 武夷山竹筏育制场 • 实联水上大楼
李克强在政府工作报告中指出，新农村建设要惠及广大农民，加强基础设施建设，要建设美丽宜居乡村。 世界反法西斯战争胜利70周年。 9月，中国纪念抗战胜利70周年大会在北京举行。 10月，中国著名药学家屠呦呦获2015年诺贝尔医学奖。 11月，东南亚国家联盟(东盟)10个成员国宣告成立共同体。 11月，巴黎发生恐怖袭击案。 12月，世界互联网大会在乌镇召开。 12月，第21届联合国气候变化大会在巴黎举行，会议签订了《巴黎协定》。	• 常青当选中国科学院院士，王建国、孟建民当选中国工程院院士。 • 伍江当选法国建筑科学院院士。 • 常青(日喀则桑珠孜宗堡保存与再生工程)获2015 亚洲建协金奖。 • 2015北京国际设计周举行，主题为“设计之都·智慧城市·产业融合”。 • 蔡永洁(北川地震纪念馆)获2015 亚洲建协金奖。 • 3月，皮亚诺“Piece by Piece”建筑展在上海开展。 • 9月，首届上海城市空间艺术季开展，主题为“城市更新”，总策展人：伍江、莫斯塔法维，建筑展策展人：李翔宁，艺术展策展人：张晴。 • 12月，“第6届深港城市\建筑双城双年展”在深圳举办。主题：城市原点，主策展人：艾伦·贝斯奇(Aaron Betsky)、阿尔弗雷多·布林伯格(Alfredo Brillembourg)和胡博特·克伦普纳(Hubert Klumpner)、刘珩(Doreen Heng Liu)。 • 梅洪元当选美国建筑师学会荣誉会士。	• 米兰世博会中国馆 • 上海中心 • 三联海边图书馆 • 哈尔滨大剧院 • 上海自然博物馆 • 北京民生现代美术馆

Landmark Nieuw-Bergen

Monadnock
Nieuw-Bergen, the Netherlands 2015

新卑尔根的地标
Monadnock 建筑事务所
荷兰，新卑尔根 2015

ı a landscape where every small village from afar can be
entified by the silhouette of its own church tower, Nieuw-
ergen lacked such a clear landmark. Recently, this need was
anted by providing the realization of a landmark building that
part of the village renewal plan. This "Landmark" is at the
re of the plan and marks the marketplace as a clear, central
llective space. The public tower offers views of the surrounding
ture reserve. It is a combination between a high, abstract tower
d a low base. It accommodates a catering facility, such as a
r or restaurant, an accessible feature that meets the need for a
ntral meeting place in the small village.
ne structure has been clad with a combination of green and
d bricks. The tower is perforated with tiny openings to let
ght shine through in the evening and then fulfils its position
s a beacon. The intention of the design is optimistic and
pproachable: it aims to be accessible and touchable. To achieve
is, the building is conceived as a small object: abstract from
ar and intimate upon a closer position. The red brick is colored
ght-green by a cement bath, making patterns emerge that scale
own the building and simultaneously enliven it.
he decorated base has the appearance of a house, in line with
e intimacy of the surrounding buildings forming the square,
n top of which two more abstract volumes are situated. Thus it
ecomes embedded in the conventional and recognizable. The
uilding is firm and clear, and plays a role as an emblem for this
illage. At the same time, it illustrates this ambition, because the
onolithic appearance in various places is put into perspective
nd shown as "bekleidung" (clothing). An emphatic ambivalence
tretches between the worlds of Rossi and Venturi.
Ionadnock has studied – and was influenced by the typology
f historical Dutch trade buildings. These odd proportioned
uildings are to be found adjacent to market squares in Dutch
nedieval cities. A quest is formulating the ambivalent status of
his building: as the building is not a church, nor a town hall, but
hoped-for representation of the collective. The planned program
f the interior will be a combination of public and commercial.

redits and Data

roject title: Landmark Nieuw-Bergen
lient: Concept-NL Project development
rogram: Tower building / watchtower
ocation: Nieuw-Bergen, The Netherlands
ompletion: May 1, 2015
rchitect: Monadnock
roject team: Sandor Naus, Job Floris, Rebecca Aguilera
tructural engineer: Bolwerk Wekers
roject manager: Monton
rbanism: LOS / Stad om Land
ain contractor: Burgtbouw

在辽阔的景观里，每座小村庄都可通过当地教堂塔楼的剪影从遥远处被轻易识别出，而新卑尔根却恰好缺少这样一个鲜明的地标。近年来，当地的乡村更新规划提出并肯定了建造地标建筑的必要性。这个"新卑尔根灯塔地标"成为更新规划的重心，将市场刻画为一个清晰的、可聚集村民的街区中心，从公共塔楼可远眺四周的自然保护区。建筑由一个高耸而抽象的塔和一个低矮的底座组成，具备一些酒吧、餐厅等餐饮设施，同时也满足了村民集会的功能需求，成为村子的中心场所。

建筑整体由红绿相间的砖体覆盖，塔身开有很多小孔，夜晚的时候光线可贯穿，使其如灯塔般标识出自己的位置。设计旨在让建筑显得乐观而亲切，即易于人们接近和触摸。因此，建筑被构想成一个小的物体：远看很抽象，走近却会感到亲密。红砖因覆有一层水泥而呈现出淡绿色，拼合出的图案适度地削弱了建筑的体量感，也使建筑显得比较活泼。

装饰的底座就像一座房子，与环绕中心广场的周边建筑所呈现的亲密感保持一致，其上是2个抽象的体块。因此，它表现出一种既常规又具有辨识度的融合感，加之建筑的坚固和明确的属性，使之得以成为村庄的象征。同时，它也在传达着这种抱负，因为不同地区整体展现出的形象就如同"衣服"一样。这种被强调的两面性在罗西和文丘里的世界里拉扯着。

设计研究了荷兰的历史贸易建筑，并受其影响。在荷兰中世纪的城市广场中，就建有这样一些比例奇怪的建筑，而这栋建筑的设计需求是塑造一种矛盾状态：既不是教堂，也不是市政厅，但又希望可以是一个它们集成的代表。同时，建筑内部兼有公共空间和商业空间的功能。

（王梦佳 译）

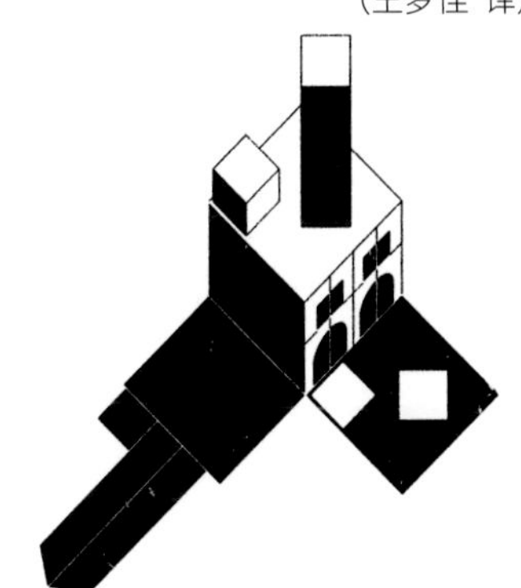

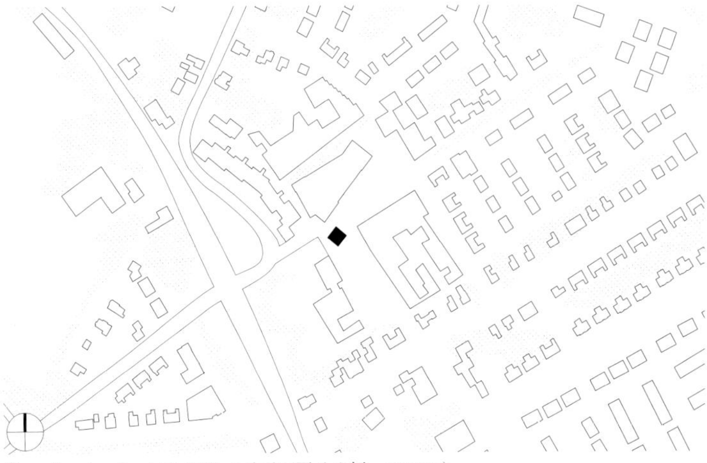

Site plan (scale: 1/3,000) / 总平面图（比例：1/3,000）

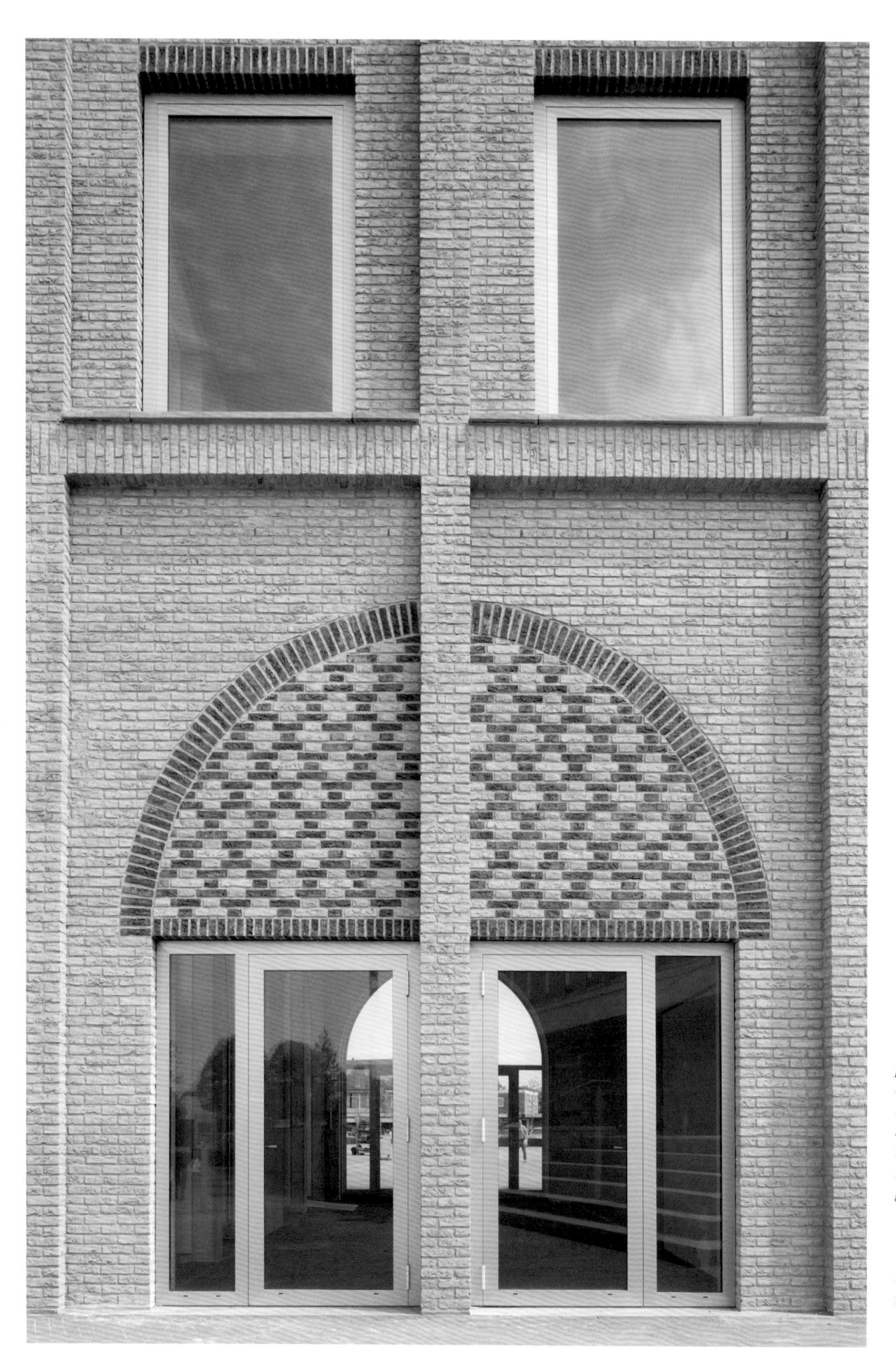

p. 164: General view from the square on the northeast. p. 165: Concept image. This page: Close-up of the southwest facade. Opposite: View inside the tower. Cement was applied to the red bricks so they appear greenish on the front. All photos on pp. 164–167 by Stijn Bollaert.

164页：从东北侧广场看的全景。165页概念分析图。本页：西南立面的近景。对页塔楼内景。将水泥抹在红砖外面使正面透出一种绿色。

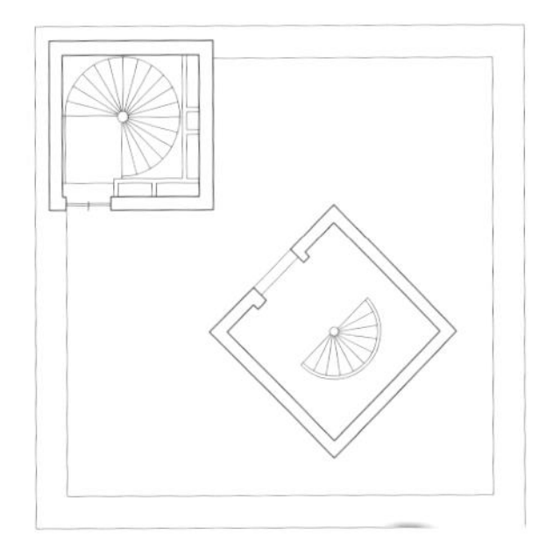

Rooftop floor plan / 屋顶平面图

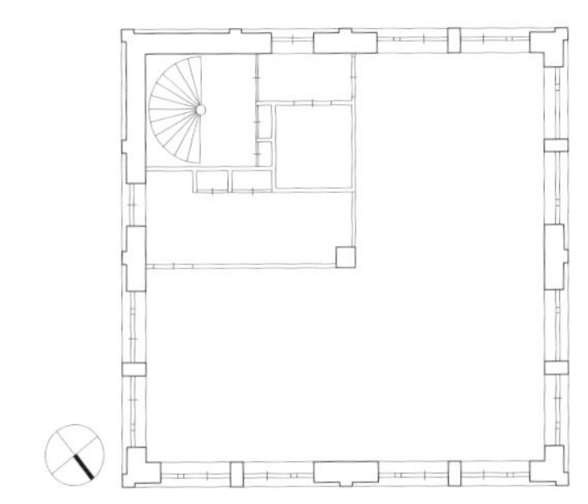

Ground floor plan (scale: 1/300) / 首层平面图（比例：1/300）

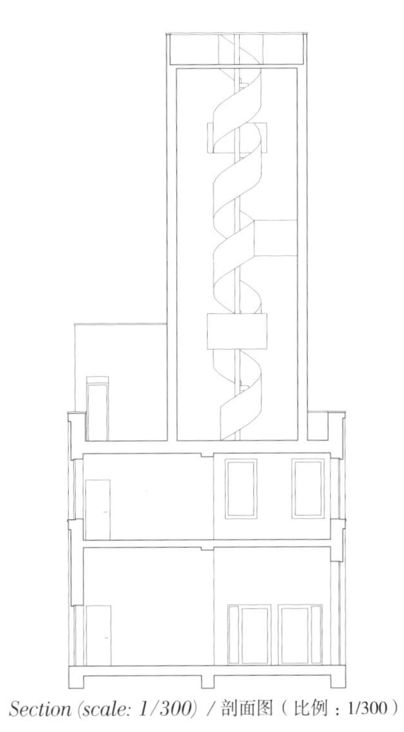

Section (scale: 1/300) / 剖面图（比例：1/300）

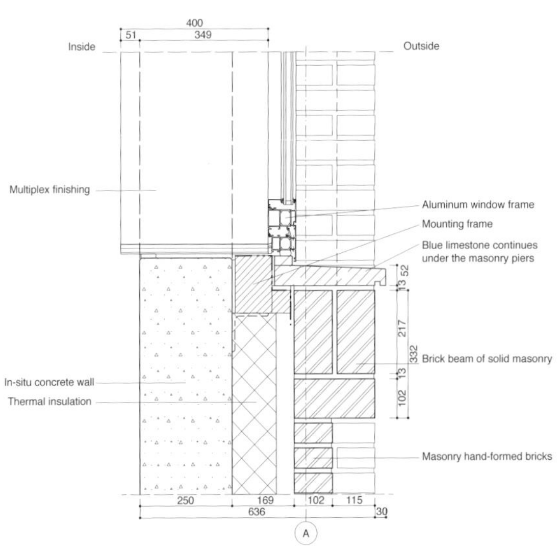

Detail wall section (scale: 1/20) / 墙体剖面细节图（比例：1/20）

Architecture and Urbanism
Chinese Edition
August, 2016 / No. 065
建筑与都市 中文版 2016年8月刊

Feature: Big + Small

特辑：Big+Small

Recent Works by BIG 收录BIG的近期作品

《建筑与都市》中文版编辑部 www.cagroup.cn
上海市大连路970号706室, 200092
TEL: +86-21-3377 3001
FAX: +86-21-3377 3006

订购方式：
同济大学出版社天猫旗舰店、微店
当当网、亚马逊、京东商城
同济大学出版社发行部
地址：上海市赤峰路2号 邮编：200092
电话：(021)65981599, 13817112028
传真：(021)65980499 联系人：朱爱民

ISBN 978-7-5608-6361-0

出版发行：同济大学出版社 www.tongjipress.com.cn / 上海市四平路 1239 号，200092 / 电话：021-6598 5622

定价：98.00 元